100最佳别墅

Top 100 Houses

佳图文化 编

華南理工大學出版社
SOUTH CHINA UNIVERSITY OF TECHNOLOGY PRESS
·广州·

图书在版编目(CIP)数据

100最佳别墅 = Top 100 houses : 英文 / 佳图文化编. —广州: 华南理工大学出版社,2013.1

ISBN 978-7-5623-3811-6

Ⅰ. ①1… Ⅱ. ①佳… Ⅲ. ①别墅-建筑设计-作品集-世界-英文 Ⅳ. ①TU241.1

中国版本图书馆CIP数据核字(2012)第244663号

100最佳别墅 Top 100 houses
佳图文化 编

出 版 人:	韩中伟
出版发行:	华南理工大学出版社
	(广州五山华南理工大学17号楼,邮编510640)
	http://www.scutpress.com.cn E—mail: scutc13@scut.edu.cn
	营销部电话: 020—87113487 87111048(传真)
策划编辑:	赖淑华
责任编辑:	王 岩 赖淑华
印 刷 者:	广州市中天彩色印刷有限公司
开 本:	889mm×1194mm 1/12 印张:38
成品尺寸:	285mm×285mm
版 次:	2013年1月第1版 2013年1月第1次印刷
定 价:	498.00元

版权所有 盗版必究   印装差错 负责调换

Preface

Collecting the latest house works around the world, the book has shown different styles and lightspots in house design. From their planning, architectural design, space design, environment design and style, it highlights each house on their concept, design, solutions and innovation with full materials including hand—drawings, planning drawings, renderings, sections, elevations, engineering drawings, detail drawings as well as high—resolution photographs. It will be an interesting book for the house designers to read and collect.

CONTENTS

001 Oceania

- 002 Moebius House
- 010 Bill's House
- 016 Letterbox House
- 020 Cloud House
- 026 Klein Bottle House

031 Europe

- 032 Dupli.Casa
- 038 K3 House
- 044 Villa Snow White
- 050 Little Black Dress
- 056 Sao Bento Residence
- 062 Villa GM
- 068 Villa Garavaglia
- 074 House CASAD
- 080 Two Family House
- 086 House F
- 090 House C
- 094 Private House in Menorca
- 100 House on Mountainside Overlooked By Castle
- 106 Fragmented House
- 110 Can Bisa House
- 114 Rehabilitation of Penthouse for Art Collectors
- 118 Atrium House
- 124 Dwelling in Etura
- 130 Villa BH
- 136 V12K0709 Piano House
- 140 House V12K0102

146	Electric Boathouse	298	Ribbon House
150	Water Villa		
156	The Balancing Barn		

303 North America

160	House at the Edge of a Forest
166	Villa Wageningen
172	Villa Frenay
178	Villa Biesvaren
184	Villa Festen

304	Nakahouse
308	Villa Allegra
312	Casey Key Guest House
318	Open House
324	Areopagus Residence
330	Echo House
336	Hill — Maheux Cottage
342	Hurteau — Miller Cottage

189 Africa

190	SGNW House
196	Glass house
204	House Moyo
212	House Pollock

347 South America

348	Pricila
354	Ribbon House
360	House L
366	Casa Maritimo
372	Residência Belvedere
378	Vila Castela Residence
384	Manifesto House
390	House Playa El Golf H4
396	Casa Playa Las Palmeras
400	Black and Red
406	Muriado
412	Psicomagia
420	Villa Ketty
426	Casa Carrara House
432	Cabo House
438	A Clear Light in the City — Devoto House
444	The Orchid

219 Asia

220	4 Connecting Boxes
224	Sentosa Cove House
228	Sentosa House
236	Villa Overlooking the Sea
242	House of Maple Leaves
246	SAZAE's House
250	Hanil Visitors' Center and Guest House
254	Floating House
262	Island House
268	Purple Hill House
274	Contemporary Bauhaus on the Carmel
280	House R — Hasharon House 1
286	House E — Haruzim House
292	Kibuts House

Moebius House

Architect: Tony Owen Partners
Location: Military Rd, Sydney, Australia
Photography: Brett Boardman

This family house faces onto views of the Sydney Opera House and Harbour Bridge. The house explores a more environmentally sensitive form of design called "micro design". Micro design utilizes parametric modeling software which can respond to very small changes to design input criteria. The unique form is a response to the requirements to maintain view and solar corridors.

Tony Owen Partners started by responding to the site with a series of movements which folded and twisted the space in order to maximize the changes of level, view opportunities and potential for connectivity to outside spaces at various ground planes. The architects created a dynamic model capable of responding to changes in these variables and allowed the models to run in real time. Then they stopped the model when the architects felt they had a model which satisfied their concerns.

The house has a fluidity of space which is a direct result of having a strong relationship with the surrounding landscape. Due to the complex geometry of this house and the need for such fine tolerances, they have to evolve a completely new system of fabrication and assembly for this house.

The architects started off designing a house, but in the end the construction process more closely resembles that of a car.

Early on it became apparent that this house would have to be detailed and documented entirely in 3 dimensions. The steel frame house is being clad in metal panels which are being pre—cut in China. The complex curving structure is like the ribs of the human body and must fir within a very slim cladding zone. The tolerances are very tight so if anything is out by even a few millimetres, the ribs will stick out from the skin. It took about 12 months to finalize the steel chasis. This involved developing the structure as a 3 dimensional model and continually checking it by inserting it onto the 3—dimensional model to make sure it fit. This model was continually checked against the computer model being prepared by the still fabricators until it was identical and all junctions were resolved.

In a traditional house the floor and walls are built first and the roof is added. The Moebius House is being

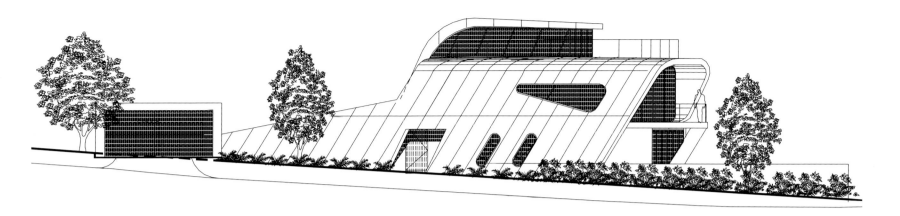

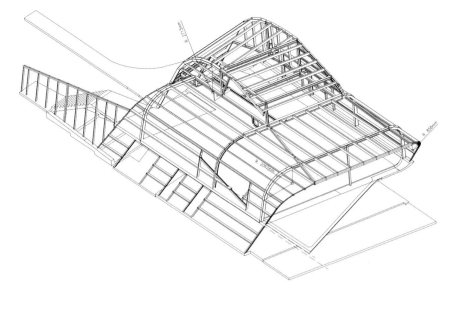

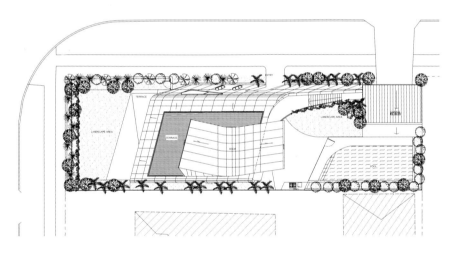
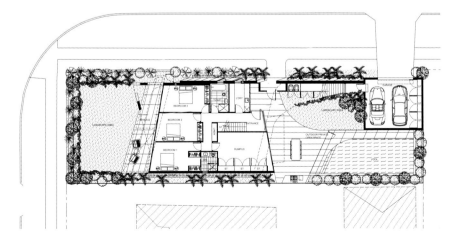

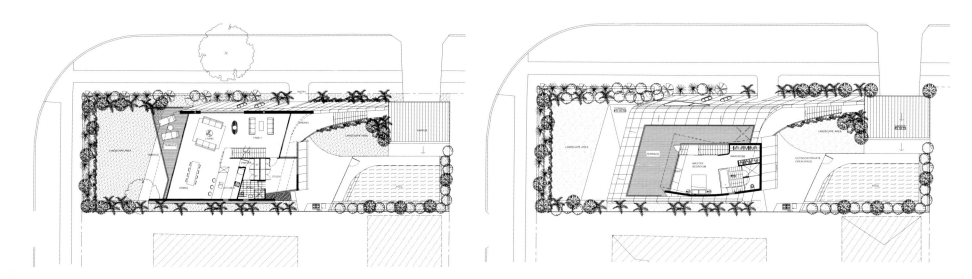

assembled around a chasis like a car would be. First the chasis is assembled on site. Then the pre—formed metal cladding panels are attached to the chasis to create a monocot shell. The house is wired and plumbed like a car, with the electrical, air conditioning and services all wired through the chasis. The kitchen even resembles a dash board.

The architects are exploring an architecture which is more responsive to the environment, and refer to this as "Elastic Design". This architecture is pliant, yet has an inherent structure and ordering principle. Elastic Architecture is an architecture that is capable of responding to all manner of changing variables. This includes spatial, programmatic, environmental and structural issues. The architects are designing spaces which expand to allow greater connectivity to the exterior environment to maximize light, air and movement flows, or retract for greater privacy and differentiation of uses. These are spaces which respond not just to program and uses, but to patterns of behaviour which change through time. The architects are envisioning dynamic buildings which respond to variations in inputs and relationships. The result is an architecture which is "future focused" in thinking, open and responsive in approach, and experimental in nature. This is an architecture which is supple and responsive, reactive to changing variables and assisted by new technology.

Bill's House

Architect: Tony Owen Partners
Location: Sydney, Australia

This house is designed for a client in Sydney's inner west. The client is a concrete contractor who plans to build the house himself. As a result the house is designed to make maximum use of concrete and solid construction. Because of the client's heritage, the design is influenced by the materials and forms of Mediterranean Architecture. The curved shell forms reflect the sails of the fishing boats from the Greek Islands. In addition the client does a lot of entertaining and wants a house which maximizs the connection to a large outdoor space.

The client's brief is to create a unique and iconic home. This presented challenges as the site is situated in a fairly homogenous suburban location. As a result, the house has been designed as a series of blocks which modulate the scale and minimize the impact of the house to the surrounding areas. It consists of a series of different internal levels, which step up progressively from the street. The house is quite solid from the street and progressively opens up to be completely open to the rear. These changes in levels create an opportunity for the strongly stepped external massing as well as the complex interplay of the stairs in the central internal spaces.

The house has an "L—shaped" configuration to maximize the solar aspect for the living spaces. There is also a central courtyard to the west which allows for light to penetrate the middle of the house and also serves to break up the massing of the facade.

A feature of the house is the large central staircase element. The original idea for this stair comes from the James Bond movie "Never Say Never Again". This stair adjoins the central courtyard so it is always bathed in light. The stair connects the various level changes in a single fluid sculptural element in dark polished concrete.

The dominant feature of the house is the curved sail—like rear white walls. These walls soften the massing and bring a lightness to the house. The walls break up the space and progressively dematerialize the house into a

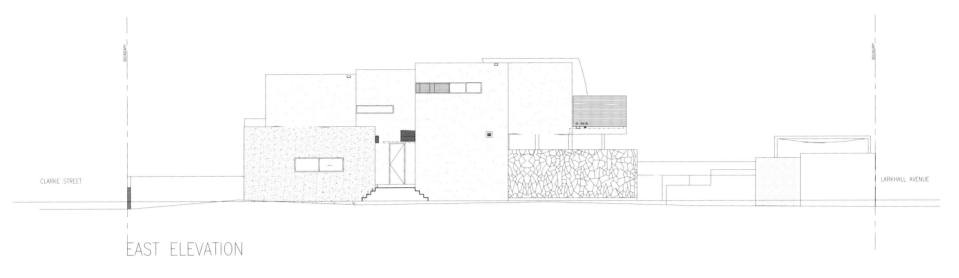

EAST ELEVATION

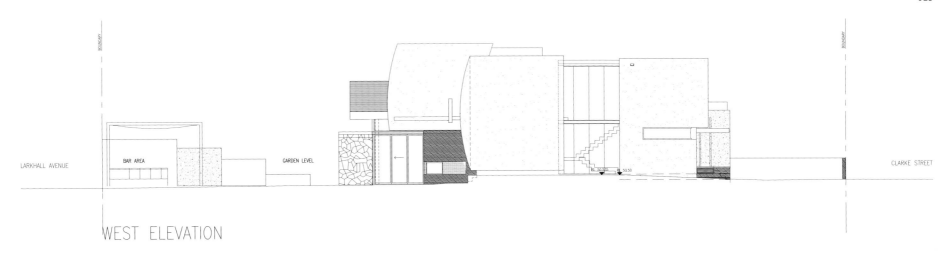

WEST ELEVATION

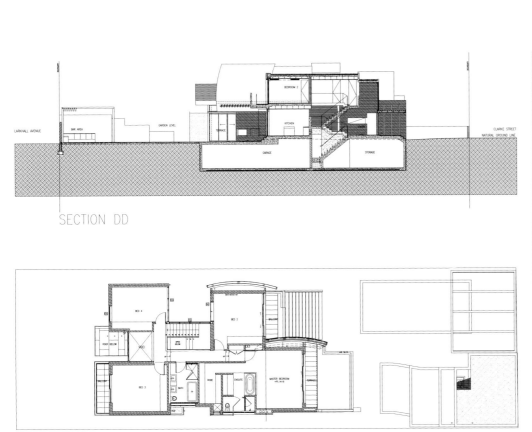

SECTION DD

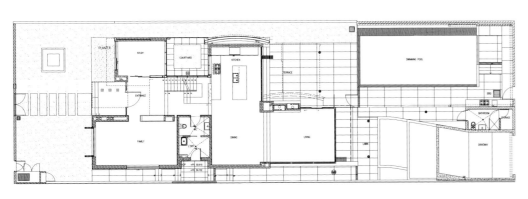

series of cantilevered vertical and horizontal planes to the rear. The kitchen and living areas of the house finally dissolve into a large outdoor room for entertaining. This space contains a glass swimming pool with a pool bar and a minimalist outdoor seating enclosure. This structure has the feel of a lounge area or bar with reclining daybeds and fabric clad structures. The use of mirrors and chandeliers enhances the luxurious lounge feel. The rear space also contains a terraced market garden which enhances the European character of the space. The living room and outdoor areas are connected by an indoor/outdoor fireplace. The spaces revolve around this element and further blur the line between inside and out.

The complex geometries of the curved structures are resolved in 3—D by computer. This house, was designed entirely using 3—D computer modelling software. This allows the architects to see the influences of the site and spatial relationships on the design. This technique is called parametric design and you can see in real time how the design changes as you adjust to changing environmental variables.

Letterbox House

Architect: McBride Charles Ryan
Location: Blairgowrie, Australia
Area: 290 m²
Photography: John Gollings

It's like a half space, half enclosed, half open. Neither in nor out—a new version of the good old Aussie verandah.

It's like a giant multi—sensory organ, the sun, the sky, the breeze and the sound and smell of the sea— When you arrive here of an evening and stand here and see the stars, no matter how still it is, you smell the sea—suck it in, it transforms you, reminds you (of what matters), it's a kind of tonic.

It's like the buildings that make you smile (not laugh).It makes people smile, a building with the smallest facade on the peninsula—the building begins as the letterbox and unfurls to become this healthy scaled verandah, to some it is an upturned boat, to others it a wave or a cliff.

It's wanted to show respect—the peninsula needs it, and the scale here was modest beach suburban—the architects wanted to respect that scale—and yet as you walk along the deck the scale sneaks up on you— before you know it you are immersed and surrounded by the scale of the house—a bit like life really.

The peninsula is the place where you suspend formality and convention for a while—It's wanted the building to do this and to remind you of that—it moves too far from architectural convention towards the other disciplines—that was the intention. It becomes ambiguous—What is it? Where is the front door? You don't need a 'front door' in a holiday house—you just find your way in.

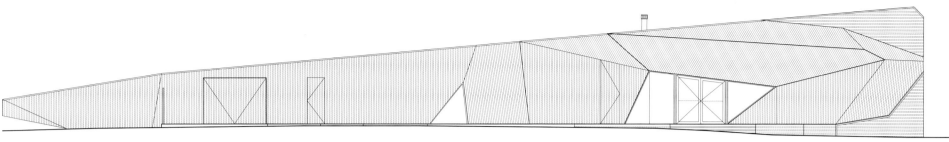

SOUTH-WEST ELEVATION

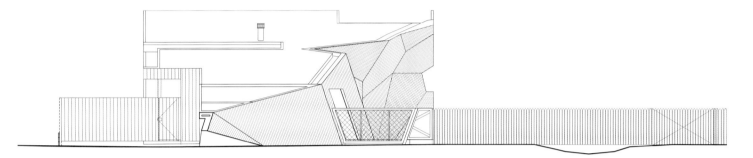

NORTH-WEST ELEVATION

Late on a sunny afternoon, when you are all salty, it is a great place to sit. the afternoon sun gives the wall a golden glow, which is echoed by the golden beer in your hand. You sit and watch the kids do what kids do—the things they forget to do when you are in the city.

The inside of this golden wall is vivid red, the support structure and the support shelves which in time will become deposits of beach memories, the much leafed books, the photos, the bric—a—brac of beach holidays and markers of the quintessential Australian family life—when that happens maybe that will then become "my space" also.

Cloud House

Architect: McBride Charles Ryan
Location: Fitzroy North VIC, Australia
Floor Area: 220 m²
Photography: John Gollings

The Cloud House is an addition and renovation to a double-fronted Edwardian house in Fitzroy North. Over the course of close to a century, this house has received several additions and modifications.

McBride Charles Ryan's work for the house is designed in three parts. This allows for a sequence of distinct and unexpected episodes, with glimpses previewing oncoming spaces and experiences as you move through the home.

The street facade has been left to demonstrate the clients' respect for the evolution of the character of the area and the modest street alteration belies the extent of the comprehensive internal renovation work. The spaces within the original structure are largely white in color, united by exotic floral hallway carpet. This journey through the space is followed by encountering a disintegrated red—colored "box". This is the kitchen, at the heart of the property, which acts as a bridge linking the major spaces. A cloud—shaped extrusion is the unexpected final space. Following the form of a child—like impression of a cloud it is a playful addition where family and friends can eat and have fun surrounded by the curved form.

The new living addition faces due south while allowing controlled north sun into the living area and providing effective cross ventilation. The form of the "cloud" conforms to setback regulations without appearing obviously determined by them. The extrusion creates a dramatic interior language where walls merge seamlessly with the floor and ceiling. The craftsmanship is remarkable throughout. it has a sense of care one typically associates with the work of a cooper or wheelwright. While the geometry is playful, the extrusion is essentially a contemporary barrel vault. It is our hope that this cloud has a "silver lining".

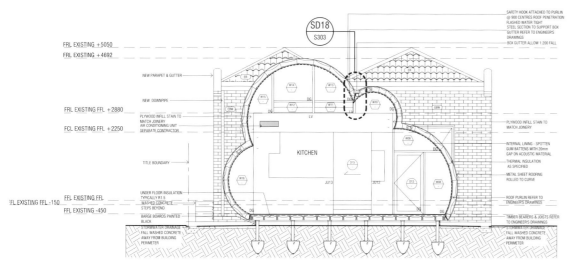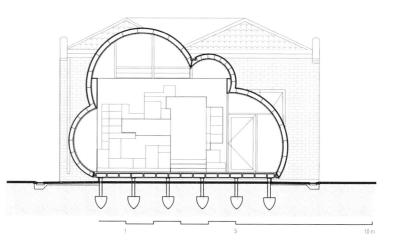

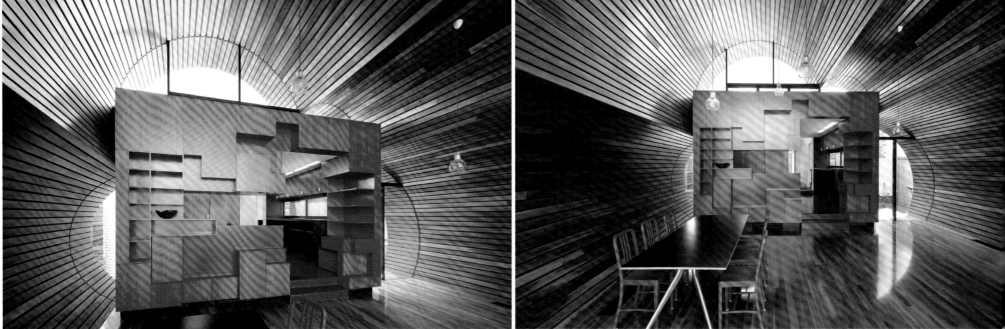

Klein Bottle House

Architect: McBride Charles Ryan
Location: Mornington Peninsula, Australia
Floor Area: 258 m²
Photography: John Gollings

The Klein bottle is a descriptive model of a surface developed by topological mathematicians. Klein bottle, mobius strips, boy surfaces, unique surfaces that while they may be distorted remain topologically the same, i.e. a donut will remain topologically a donut if you twist and distort it, it will only change topologically if it is cut.

The surfaces that mathematicians have developed hold intrigue for architects as they hold a promise of new spatial relationships and configurations. Technology (CAD) has played an important part in all this, it is now more possible to efficiently describe more complex shapes and spaces and communicate these to the build. Previously the more orthogonal means of communication plans, sections and elevations naturally encourage buildings which are more easily described in these terms, i.e. Boxes.

This holiday house is situated on the Mornington Peninsula 1.5 hrs drive from Melbourne. It is located within the tee—tree on the sand dunes, a short distance from the wild 16 beach. From the outset MCR want a building that nestled within the tree line. That talked about journey and the playfulness of holiday time. What began as a spiral or shell like building developed into a more complex spiral, the Klein bottle. MCR are keen to be topologically true to the Klein bottle but it has to function as a home. An origami version of the bottle would be achievable and hold some ironic fascination. (The resulting FC version also has a comforting relationship to the tradition of the Aussie cement sheet beach house).

The building 1950's is also within that tradition of the use of an experimental geometry that could be adapted to more suitably meet contemporary needs, and desires. In that sense it is within the heroic tradition of invigorating the very nature of the home, most notable of this tradition would be the great experimental heroic houses by Melbourne architects in the 1950's (McIntyre and Boyd in particular).

The house revolves around a central courtyard, a grand regal stair connecting all the levels. There is a sense of both being near and far to all occupants.

Its endless, curling shell—like quality particularly in the tee tree brings about a comforting togetherness.

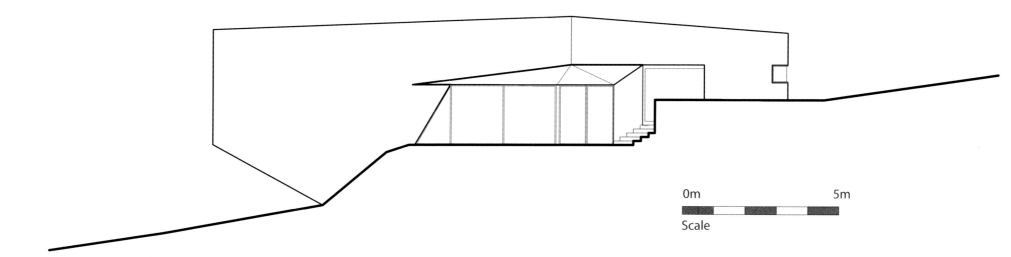

Dupli.Casa

Architect: J. Mayer H. Architects
Location: Ludwigsburg, Germany
Site Area: 6,900 m²
Photography: David Franck

The geometry of the building is based on the footprint of the house that previously was located on the site. Originally built in 1984 and with many extensions and modifications since then, the new building echoes the "family archaeology" by duplication and rotation. Lifted up, it creates a semi—public space on ground level between two layers of discretion. The skin of the villa performs a sophisticated connection between inside and outside and offers spectacular views onto the old town of Marbach and the German national literature archive on the other side of the Neckar valley.

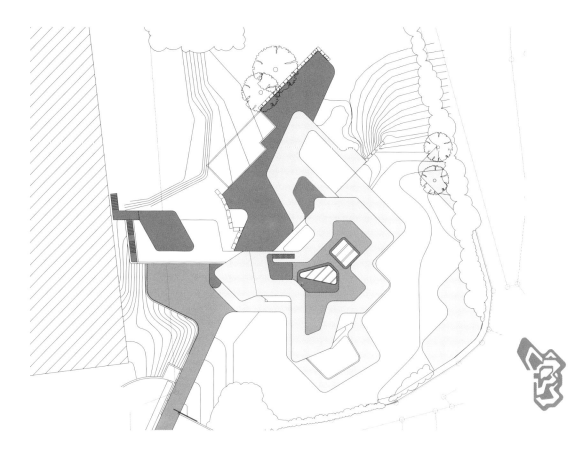

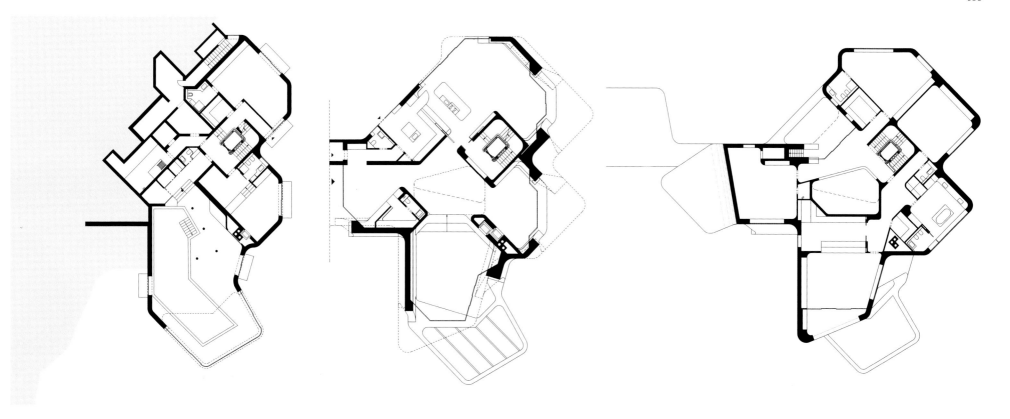

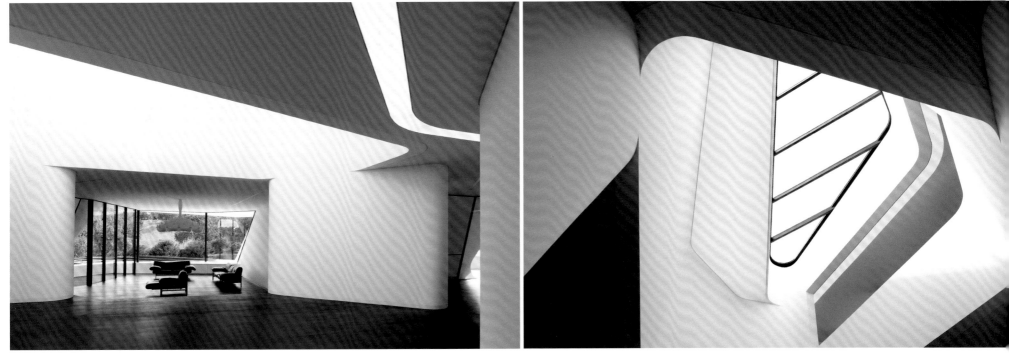

K3 House

Architect: BSA
Location: Vaucluse, Provence, France
Photography: Karl Beath

This dramatic renovation centres around a spacious internal courtyard defined by natural rock face and lush vegetation. Large sliding glass doors in the main living area enable a seamless flow between inside and outside.

The living areas also have the added benefit of glazing on the north facade which opens up the house to the view.

The master suite pavilion, perched on the highest portion of the rock face, has been designed as a sanctuary for the parents, whilst maintaining a bird's eye view over the living areas.

040 | Europe

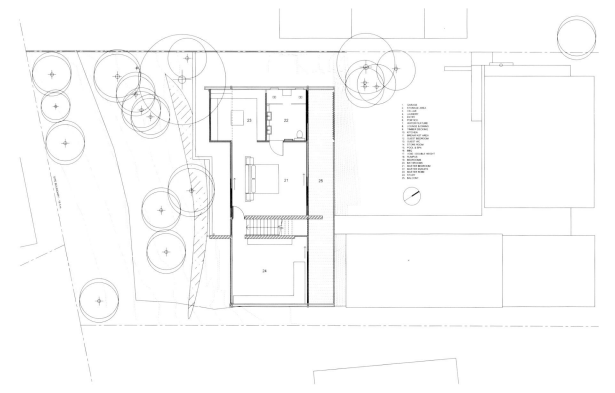
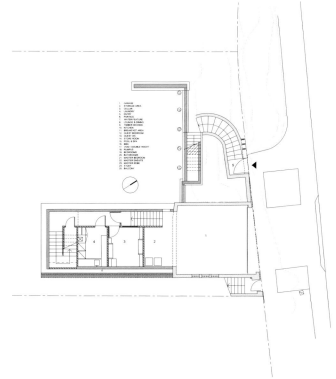

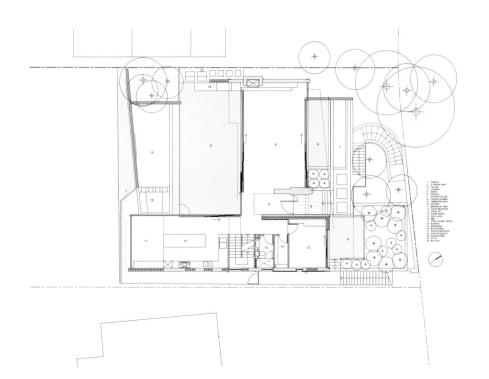
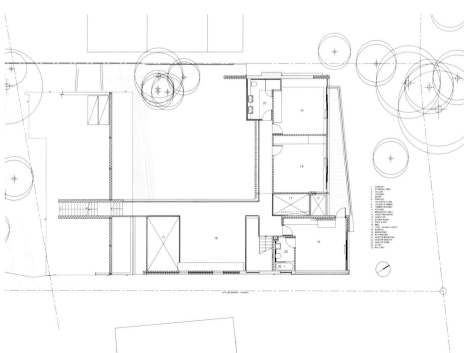

Villa Snow White

Architect: Helin & Co Architects
Location: Espoo, Finland
Area: 590 m²
Photography: Mandi Tuominen

The building locates on the top of a hill facing south, towards the archipelago spreading along the Espoo coast.

An obsolete old villa was cleared out off the property before the construction of the new building. The neighbors of the property are a ten year old flat roof detached house and a villa from the 50s.

The old trees of the property play a big role in the designing process. The openings of the interiors are set by their location.

Woodnotes Design and the authentic minimalism of their products set the direction for the visual style of the building. It is the client's wish that the building would display the same feel of authentic materials that is also found upon Ritva Puotila's textile art.

The building serves two functions: it works both as the client's family home and as a showroom for Woodnotes.

The main entrance locates through a closed outdoor atrium, facing the West towards the evening sun. The atrium can also be entered from the kitchen and via a bridge and a staircase from the sauna.

The spaces live with the Northern light and adapt the white surfaces to the scenery in which it locates all the way to the distant horizon of the sea.

The concrete structure of the building is casted on the location over steel pillars and cladded with snow white plastered tiles. The surfaces of the interiors are light and the floors are mostly solid oak and limestone.

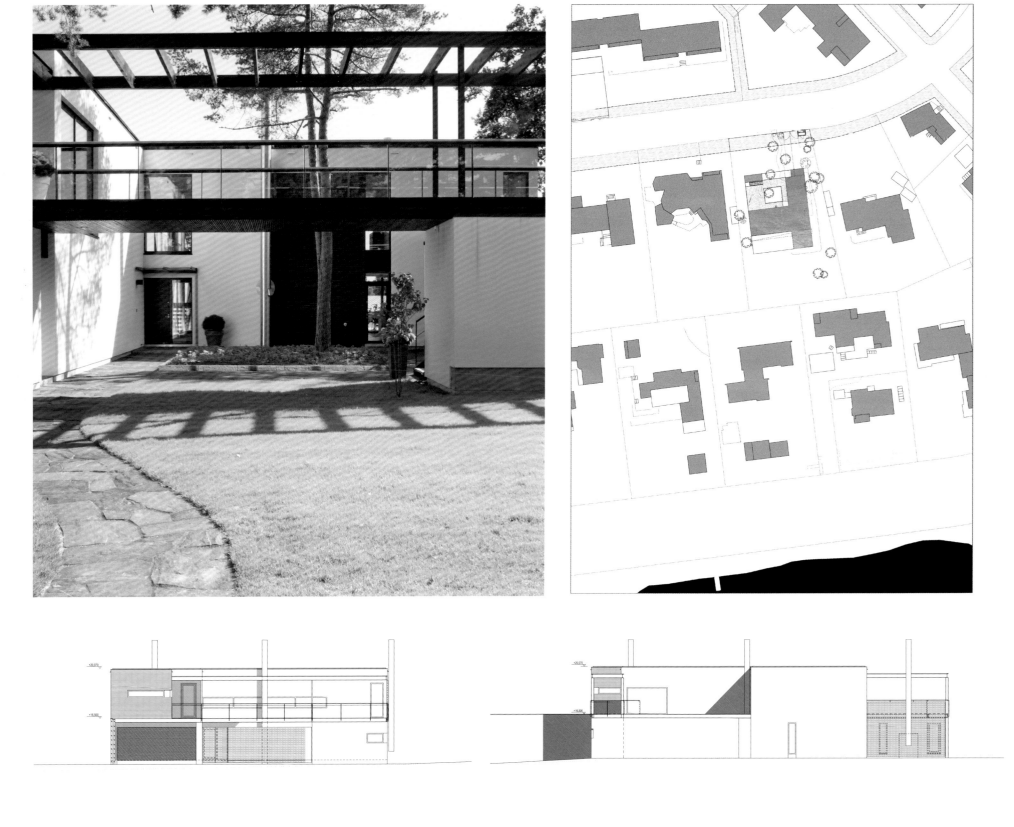

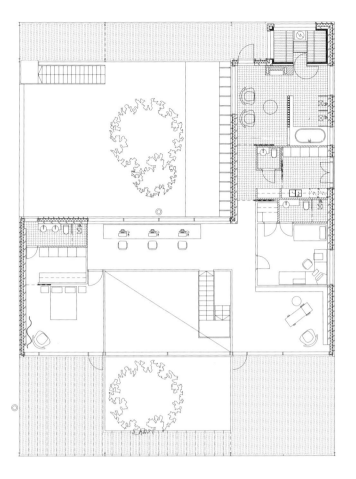
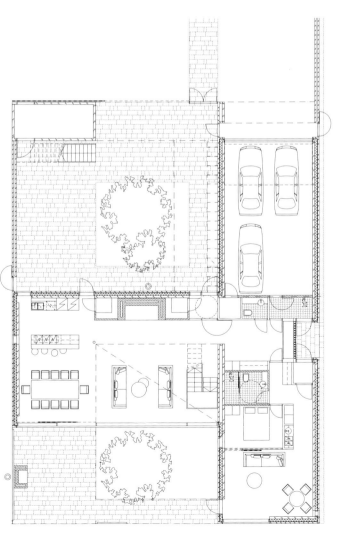

Little Black Dress

Architect: AllesWirdGut Architektur ZT GmbH
Location: Wien, Austria
Gross Floor Area: 510m²

Vienna's 14th district, at the fringe of the Vienna Woods: on a site sloping steeply in North—South direction sits the very compact—looking, single—family house.

The first impression is reinforced by the matte black skin of the building. Inside, however, it offers its occupants a spacious and varied living environment on seven staggered half—story levels. At each level, the house opens differently to the outside world. A hillside cut and a court below grade provide for natural lighting of the access area and multi—purpose room. The basement becomes a full—blown living area.

Above it, the main residential levels are nestled against the slope, separated from the garden only by an all around—strip of windows which allows looking and stepping out in every direction. In the center of the living area, there is a core of unclad raw reinforced concrete with the kitchen and hearth. It supports and provides discreet access to the private retreat spaces that form the roof area. Here, the parent's and children's bedrooms are situated at separate levels, each with a loggia that affords absolute privacy. The roof area is crowned by the bathroom and sauna level with a terrace offering a panoramic view of the surrounding Vienna Woods.

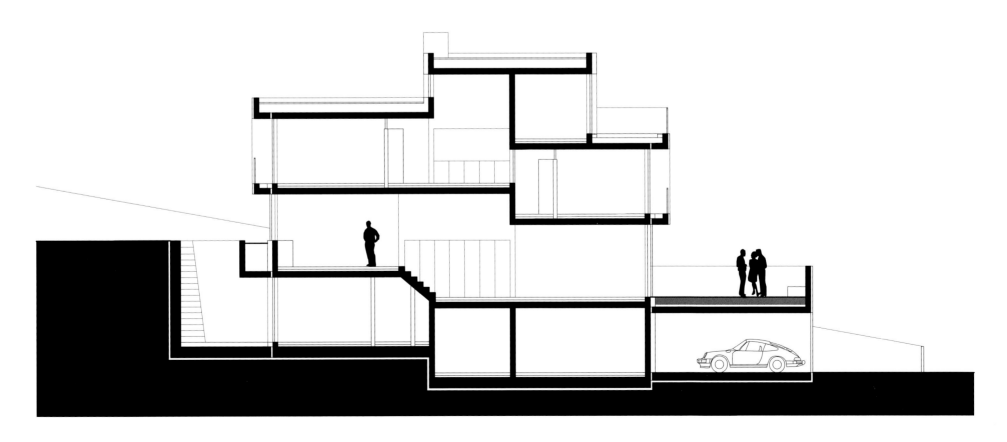

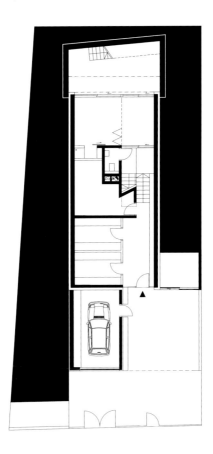 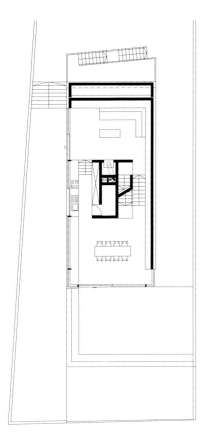 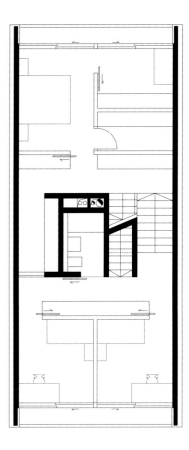 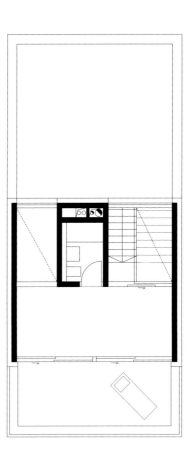

Sao Bento Residence

Architect: Anastasia Arquitetos
Location: Sao Bento, Portugal
Site Area: 840 m²

The residence is situated in a 15 m x 56 m sloping urban site. The land has a direction east / west, with the best view to the east, and the entrance to the west.

The idea is to get the most possible exposure of the intimate area, on the second floor, to the morning sun, and also to the best sight. The solution is to extend the floor diagonally on the lower floor, creating a terrace for the intimate area, so that all bedrooms face east, which would not be possible if all were side by side.

The west front, facing the street, on the need for privacy (since the Architect's intention is to keep the rail leaked) and sun protection has a few openings and the main opening is the entrance of the house. As the site is a steep slope, the ground level is 3 meters below the street greider, then the main entrance is moved back 17 meters to the alignment of the site so we could get a pleasant access and the house could have a better ratio in addition to a more generous landscaping.

Established the party the form came naturally with the curve of the second floor as part of its striking architecture. The inclination of the wall reinforces the architectural intent with the curved shape. On a street where all the residences are closed by walls on the border of the sidewalks, the architects created a moment of visual refreshment for the pedestrians.

The spatial distribution is designed to take advantage of the terrain. Thus the space under the house is used for parking of vehicles and laundry. At the ground level is the social and service's areas (living, dining, kitchen, service area, maid's room—being the living area and the kitchen integrated with the balcony and the pool), and also there is a guest room. On the second floor are the bedrooms and an intimate lounge.

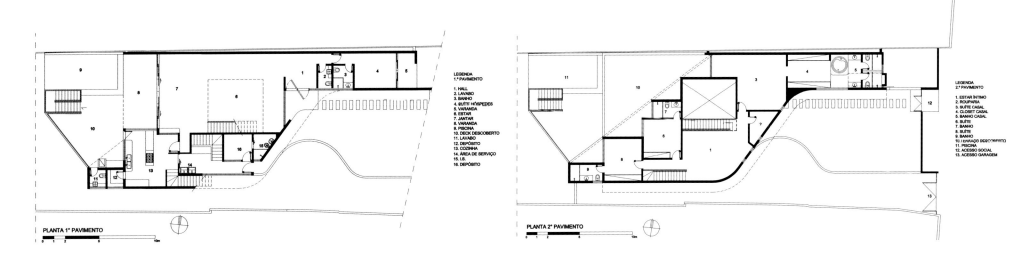

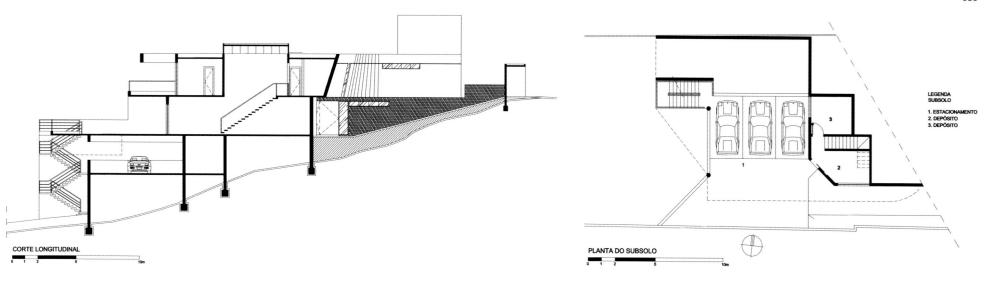

CORTE LONGITUDINAL

PLANTA DO SUBSOLO

LEGENDA
SUBSOLO
1. ESTACIONAMENTO
2. DEPÓSITO
3. DEPÓSITO

The structure of the house is concrete. To tilt the outer wall, balancing the cover slab 80cm over the slab bottom the architects tie the bricks to each 1.50 meters with concrete pillars, since it would be very expensive and also would overload the structure of the house if the wall are made of concrete.

The lighting is a crucial factor in the project. Due to the greater length of the site has neighbors on both sides, the architects have the two smaller fronts to illuminate the house, and the west facade needs to be closed because of the need for privacy. Thus a zenithal lighting is created over the standing double duty of the room to illuminate the core of the room.

The biggest challenge of the house is the site itself, and its strong ascent, at its greatest extent confined by buildings on both sides. In the end, the result shows a strong and respectful presence in the context which it appears: a very nice house where the shape does not compromise internal space, on the contrary, this is the consequence.

Villa GM

Location: Marina di Ragusa, Sicily, Italy
Architects: Architrend Architecture
Gaetano Manganello & Carmelo Tumino
Collaborators: Patrizia Anfuso, Marco Garfi
Area: 1,250 m²
Photography: Moreno Maggi

This villa, like a garden pavilion hung with a spectacular view of the sea, is part of a complex of houses located in Marina di Ragusa, the seafaring village of Ragusa, on a plot of land with beautiful views overlooking the Mediterranean and a stretch of coast in the direction of the island of Malta, distant about sixty miles off, and that in a bright day you can see clearly. The design of the villa derives from the influence exercised by the program of the Case Study Houses (CSH) implemented in the '50s by John Entenza and the magazine he founded "Art & Architecture". The house is more representative of the program is certainly the case study houses of Pierre Koenig's Stahl House, masterfully photographed by Julius Shulman, became an icon of American lifestyle in the famous photo of the living room of the house with the background on the amazing night view of Los Angeles. The position of the batch of the project and the cultural similarities with that program became the essence of contemporary absolutely present after more than fifty years, has determined the main choices that affect the architecture of the house.

The villa has an L-shaped ground plan shape, is set around a large swimming pool with sun terrace paved with planks of larch treated with a white primer. The continuity of the interior of the living room and is secured by a glass wall that continues to spread around the perimeter of the house facing the sea view. Compared to the garden the house is almost suspended, because a continuous and smooth edge, detached from the ground, surrounds the house, determine the line of coverage, the line connects with the base, is defined by vertical sidewalls. Two walls demarcating the inlet side and the opposite border of the pool are independent of the structure and connected with it through a high window and continuous thought of as individual plates that slide, too detached from the line of soil and structure. Architecture is dry and clear, made so well by the economy of the materials used, steel and wood frame, glass for the side walls and cement floors for both internal and external. The only element of disturbance, at a scale consisting of a thin folded sheet which is held at a red carpet to mark the entrance. The garden was designed as a collection of Mediterranean plants with the edge of the sea area defined by a sinuously to the cacti,

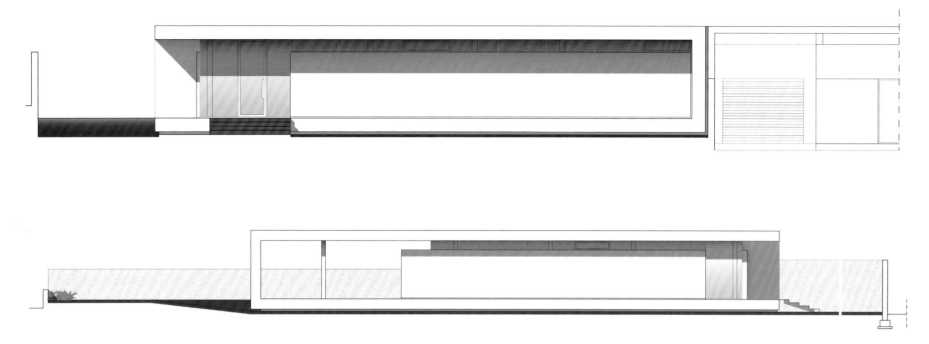

palm trees here and there are some underlying organic forms in which the white gravel of the materials are clearly delineate the Green lawns. The interiors are essential ports are designed as full-height panels of the same thickness of the partition, and then back to the wall on both the internal and the external room hallway are white lacquered opaque white as the walls the house, white as the wooden beams of the roof. The white fluid that surrounds the dimension of the interiors, white is also the kitchen island facing out to sea and pool. Gray are the seats of living a light gray shading into the white furniture and the gray cement floor. The dining table is the Saarinen Tulip, the chairs are the Series 7 by Arne Jacobsen, while the light above the table is a ball of light in glass such as the Stahl house. The charm of the Stahl house determined the character of the villa, its planimetric shape, its design philosophy with the choice of using the exposed structure. The whole house is a tribute to its architect Pierre Koeing, perhaps the most brilliant architects with Craig Ellwood Americans who have given it a great contemporary American and world architecture.

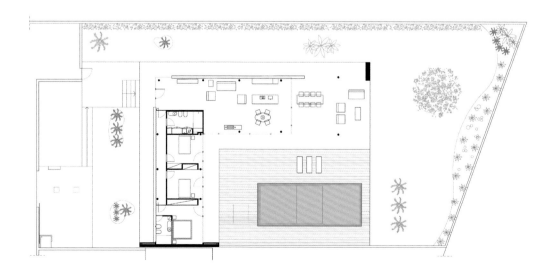

Villa Garavaglia

Architect: Buratti+Battiston Architects
Location: Mesero (Mi), Italy
Photography: Marcello Mariana

This project of Villa Garavaglia in Mesaro (Milano) is a part of a long research about the house and the domestic interior that Buratti+Battiston Architects worked out in these years with many projects of houses in suburban areas of Milan.

The relation between inner and outside spaces and the research for a design process that can link space articulation inside with volumes composition outside are the main features of this project.

This is a house for a family of three, parents and a daughter, where day and night areas are placed in the ground floor, and a studio with guest room is in the upper level just under the roof.

The one—lean roof, the main compositional and typological element of the project, is folded and cut to design interior and external spaces of the house: a big double—height living room is placed beside external patio covered and open to the garden.

Buratti+Battiston Architects designed also interior and furniture, and in all the rooms of the house the quality of the space is gained by clear and severe finishing exalted by natural light.

The palette of the materials used in the exterior is composed by white plaster for walls, teak wood for windows, gray stone for floor and copper for the roof.

This selection of materials characterized the composition of volumes and many details "super—designed", as the one of the drain gutter, qualifying the whole architecture.

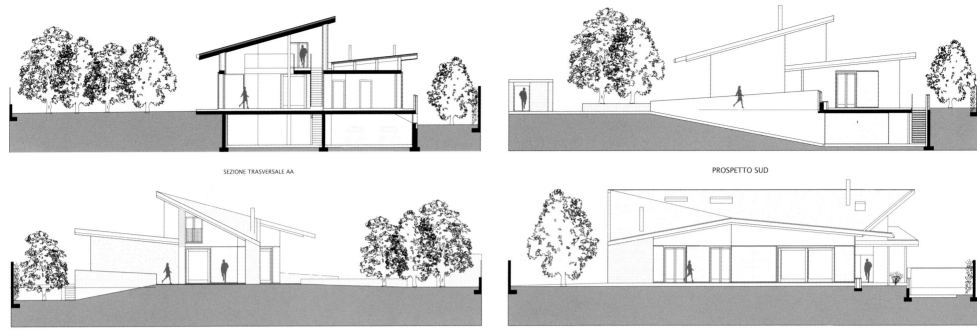

SEZIONE TRASVERSALE AA

PROSPETTO SUD

PROSPETTO NORD

PROSPETTO OVEST

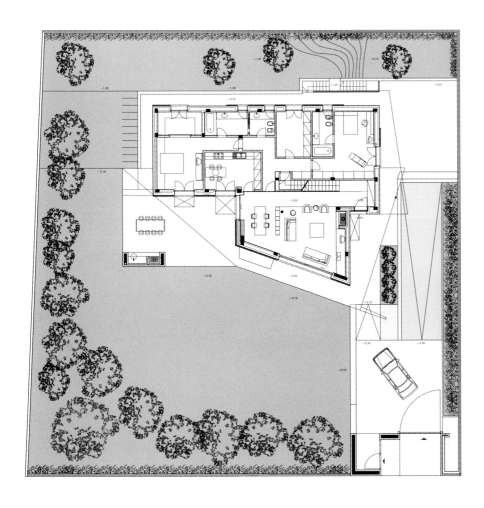
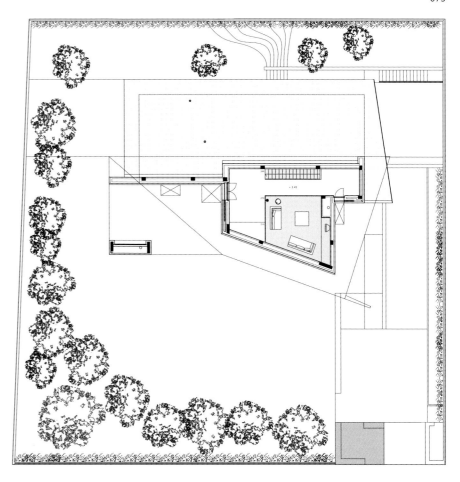

House CASAD

Architect: Damilano Studio Architects
Location: Cuneo, Piemonte, Italy
Surface Building: 355 m²
Photography: Andrea Martiradonna

The house is on a gently sloping terrain. That give the pretext to develop a house on two levels, connected by a mezzanine floor.

The entrance to the house has a curious glimpse of the dining area, located south—east of the Appendix. The living room opens onto a large window on the porch and garden, raised above the share of the pool. Large outdoor terraces that follow the natural slope of the land, linking the summer lunch at the poolside and the lawn even located at the lower level, overlook the spa area.

The mezzanine level of the house is the double—height library, which separates the living from the sleeping area. Given the passion for reading of the commissioners, the library is conceived as the hub of the home, relax in direct contact with water, vibrant and evocative. Below the window of another residence illuminates the cutting of children's play area. Fossil stone walls, enclosing the library enter the residence without interruption.

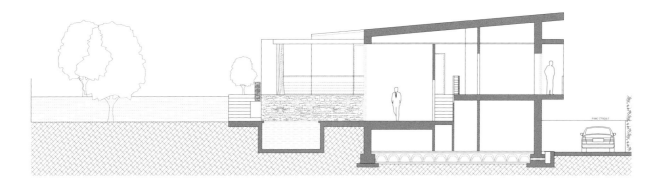

SEZIONE A-A'/ A-A' SECTION

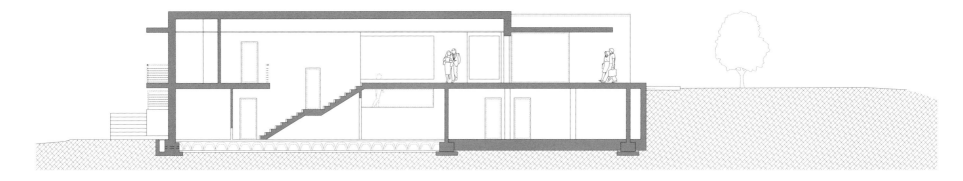

SEZIONE B-B'/ B-B' SECTION

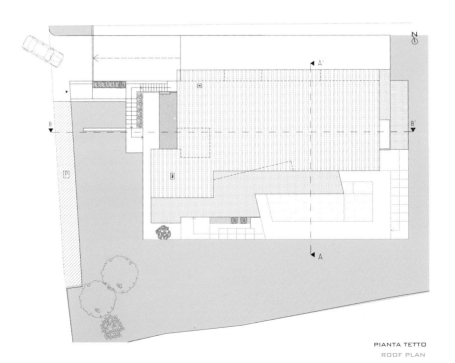

PIANTA TETTO
ROOF PLAN

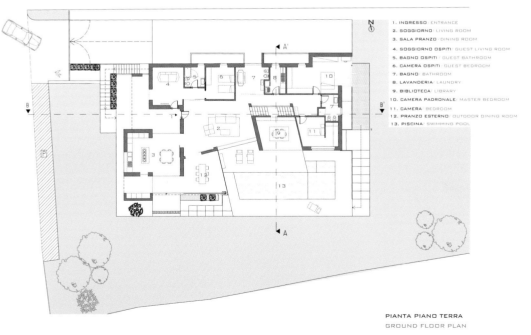

1. INGRESSO: ENTRANCE
2. SOGGIORNO: LIVING ROOM
3. SALA PRANZO: DINING ROOM
4. SOGGIORNO OSPITI: GUEST LIVING ROOM
5. BAGNO OSPITI: GUEST BATHROOM
6. CAMERA OSPITI: GUEST BEDROOM
7. BAGNO: BATHROOM
8. LAVANDERIA: LAUNDRY
9. BIBLIOTECA: LIBRARY
10. CAMERA PADRONALE: MASTER BEDROOM
11. CAMERA: BEDROOM
12. PRANZO ESTERNO: OUTDOOR DINING ROOM
13. PISCINA: SWIMMING POOL

PIANTA PIANO TERRA
GROUND FLOOR PLAN

Two Family House

Architect: Monovolume Architecture
Location: Kaltern, Italy
Built Area: 600 m²
Photography: René Riller

Giacomuzzi house is a duplex building. It is conceived as a highly compact body to fulfill energy issues and it fits into the sloping terrain to exploit the landscape. The house is oriented towards the path of the sun and overlooks a stunning sight. The site is located on the outskirts of the village Caldaro sulla strada del vino, where between the house and the lake Caldaro there are just fields and vineyards.

To show up this wonderful view, a continuous wall closes three sides of the building, whereas the south facade features wide windows and balconies. The living room and bedrooms face to south, the other rooms to northwest. This type of duplex house is the modern interpretation of the traditional pitched roof houses, to exploit the solar energy. Following the path of the natural slope, the position of the first and second floor is shifted backward from the lower floor.

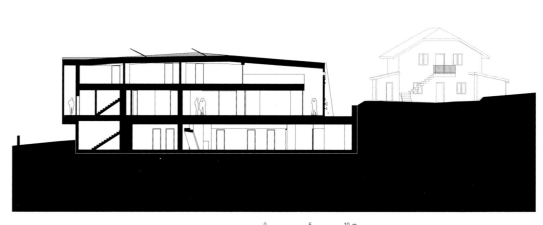

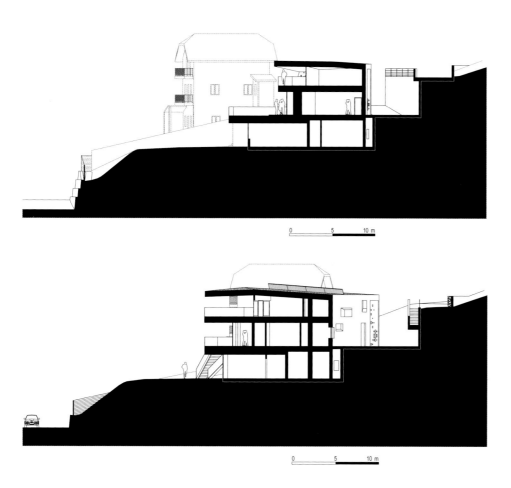

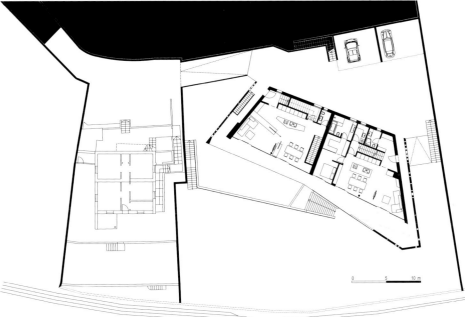

House F

Architect: RTA—Office
Location: Barcelona, Spain
Photography: Casa Ferran

This house designed by Santiago Parramón (RTA—Office) has as main characteristic its location in a land with a high incline and a stream that crosses it. A rocky orography and the preservation of the most of the trees are the guidelines of the project.

The objective of Santiago Parramón is to take up minimally the site in order to give freedom to the stream, so they let you hear the sound of the water, the smell of vegetation, the textures... All of this comes on of top the emotional aspect to the architects wants.

Finally the architects decided preserve the site intact to lean on it an object, a big volume which rests in a minimum surface. The house only takes contact with the land through a central nucleus with small dimensions. It was an archaeological work where the architects worked as a surgeon in order to minimize the impact on the rocky land.

Inside the house, in the central nucleus, there is a stair which extends to the covered, where two big mean beams throw themselves to the built sides. The forged of the interior floors lean on the central nucleus and they throw themselves to find the struts that hang from the mean beams. The house is hanging.

Santiago Parramón seeks that the project is understood as a unique object that is the reason why the entire program and the interaction between it and the exterior is produced inside. The objective is to create a space inside the object, a carving object which has been penetrated vertically, horizontally and transversally and, at the same time, those penetrations allow to communicate with the exterior. These tunnels are completely covered (floors, walls and ceilings) with pale—blue quartzite.

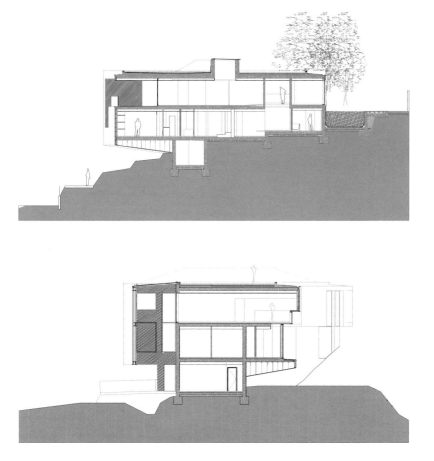

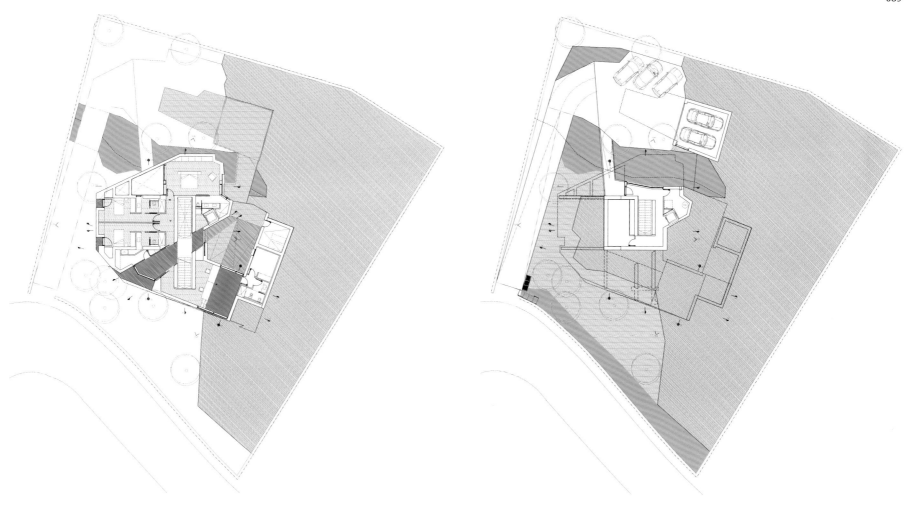

Interior and exterior are white and all the interactions between them are solved with the same material, pale—blue quartzite whose characteristics give a soft reflex that makes possible that the nature enter inside. Only white and pale—blue: body and drilling.

By this way, the landscape can be seen through a glazed surface at the end of the corridor and also reflected in a kaleidoscopic way through inside. One of these vertical corridors allows seeing the stream from the main room. Interior is completely coated of natural light having continuous and wide spaces. Magnitude can be felt in its full dimension.

The pale color of the outside of the house seeks a contrast between the object and the vegetation. As soon as the vegetation and the trees grow up, the architects will reach its objective of discovering the house hidden behind the trees.

It is a Studio—house. The program is disposed in three levels. The main floor is located on the top level and just bellow there is a room's floor. In the last level there are the complementary services.

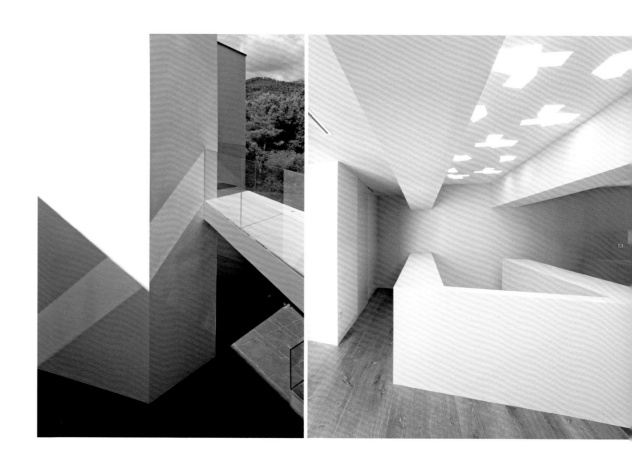

House C

Architect: RTA—Office
Location: Barcelona, Spain

Geometry, skin, cuts that let you discover the dermis. Space, three—dimensional continuity. There are no elevations, no plans, or sections. This is space within a solid object, a block of black basalt in which the architects make a number of penetrations as in a quarry, mine tunnels and corridors that communicate with the outside, with the light of the morning, of noon and the darkness of the night.

An emotional design, the indivisible cross—sectional view of the object. Dark skin, pale dermis. Contrast.

Cuts and cracks, reinforcing the change of scale, working for the presence. A unitary object that proposes its maximum dimension. The complexity of the site works in the architect's favor: the closed outer perimeter, walled, stoney; inside the glass folds open.

The architects traverse the wall. It's the north face of the building and there the architects place the access: a spectacle of natural light, reflections, transparencies, an explosion of multiplying images. A kaleidoscopic space, beauty, image, observation. All at the same time thanks to the effects produced by the transparent crystals of different sizes and angles. Everything impacts on the building. Nature enters the interior through these cuts and segments the different rooms of the house. At dusk, transparency overrides limits: the light of day is now projected from inside the house to the outdoors.

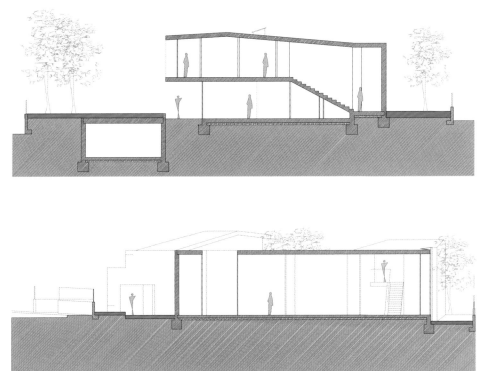
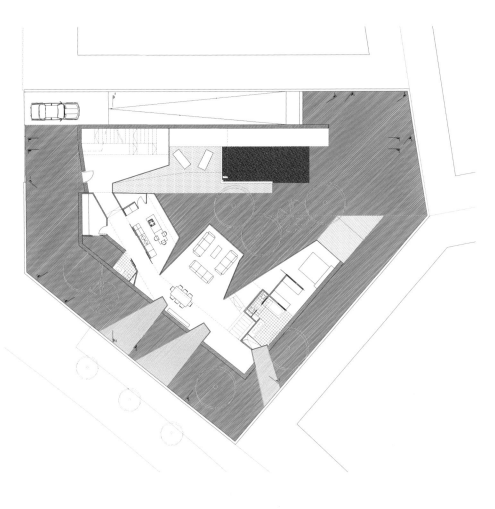

Private House in Menorca

Architect: Pablo Serrano Elorduy
Location: Ciutadella de Menorca, Spain
Area: 443 m²

Being a summer house, the main idea is not only creating the interior spaces of the house, but distributing all outer space. The interior spaces seek good relations with the outside world, colonizing their surroundings and their views.

Inspired by the typical Menorca "tanca", stone walls divisions of the realm. The plot is organized from a space frame, fully passable, based on a trace orthogonal, combining floors, platforms, water, trees, plants, tanca, pergolas, walls and the house itself. By combining these elements the architects are encountering this approach in which each piece is delimited and acquires its own identity and use within a harmonious whole. The diversity of outdoor stays provides the site a space balanced richness.

The house is situated in the center of the outer solar stays divided in two, front and rear. The hall of the house with two large openings on each side operates as a mixed external—internal transition. Falls outside the pavement causes a passage that connects the back yard with the front porch.

Based on typical lattices "menorquinas" designed as sliding wooden slats fixed set a filter to the outside, they act as sunscreens, giving privacy and multiplying the usability. They combine the vertical slats of the east and west facades with horizontal south facade. Most of these openings are floor to ceiling, allowing for greater continuity to the outside and making the most magnificent views of the site.

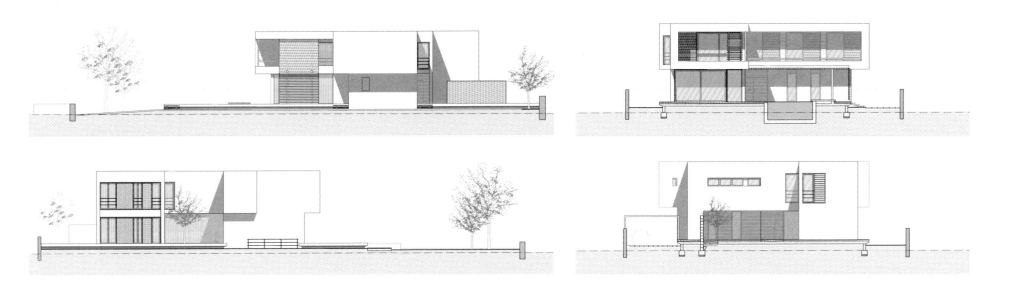

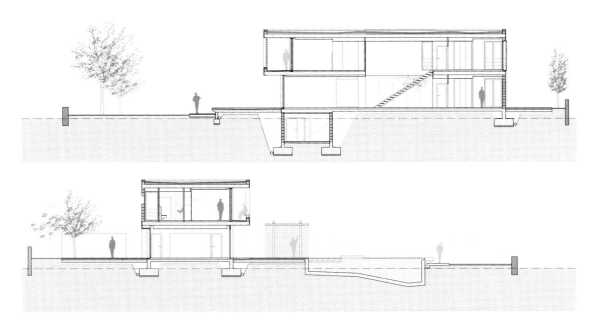

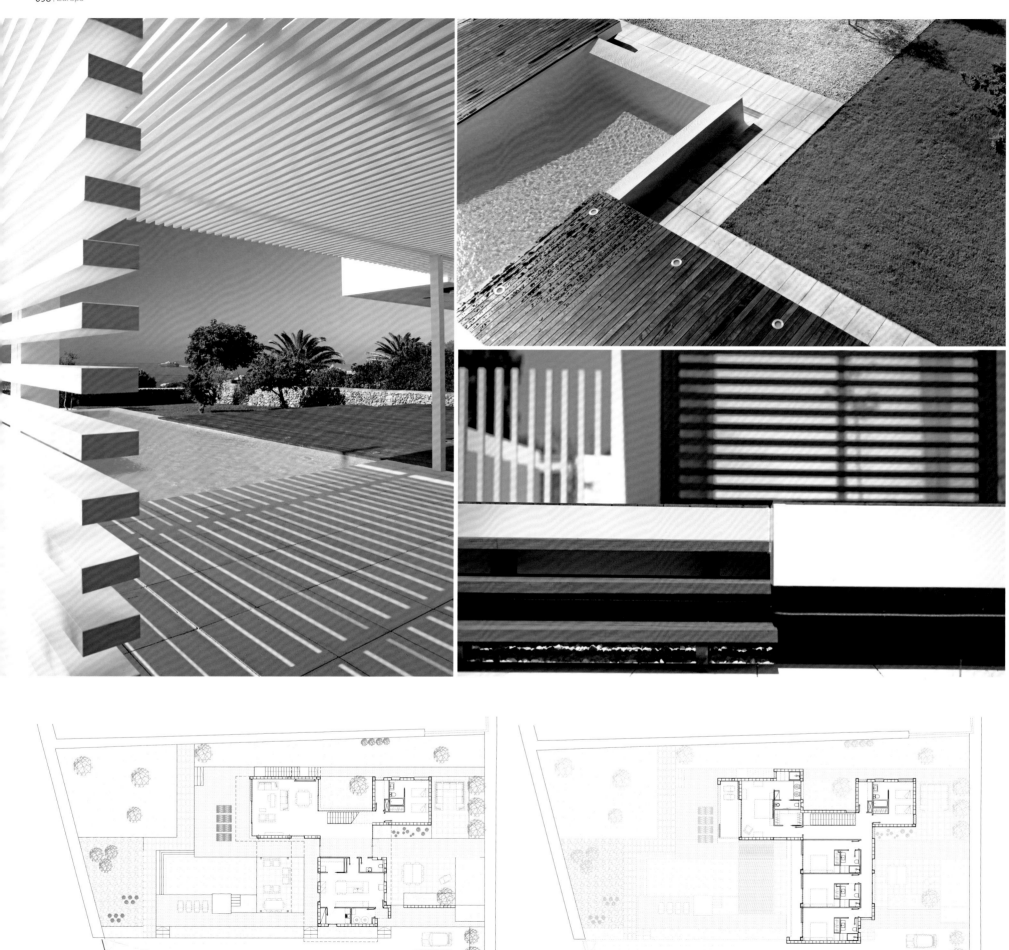

House on Mountainside Overlooked By Castle

Architect: Fran Silvestre Arquitectos
Location: Ayora, Valencia, Spain
Site Area: 477,06 m²
Photography: Fernando Alda

The building is located in a landscape of unique beauty, the result of a natural and evident growth. The mountain, topped by a castle, is covered by a blanket housing through a system of aggregation by simple juxtaposition of pieces generated fragmented target tissue that adapts to the topography.

The project proposes to integrate into the environment, respecting their strategies of adaptation to the environment and materials away from the mimesis that would lead to misleading historicism, and showing the time constructively to meet the requirements of the "new people". In this way the house is conceived as a piece placed on the ground, joining in the gap. A piece built on the same white lime, the same primacy of the massif on the opening, which takes the edge of the site to have their holes and integrated into the fragmentation of the environment.

The indoor space is divided by the void that is the core of communication cut parallel disposition of the mountain without touching it. On the ground floor are the garage and cellar, on a volume it has two floors with four rooms. Two of them, the rooms at the intermediate level are open to the private street, the other two on the upper level overlook above the houses opposite, the Valley of Ayora. One of them, the study is opened in turn to the central double height, incorporating it into their space. Across the gap, and on the mountain, are the areas facing the garden day illuminated by light reflected on the south slope of the castle oxidized.

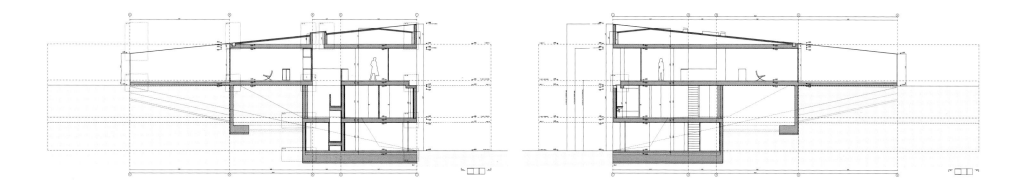

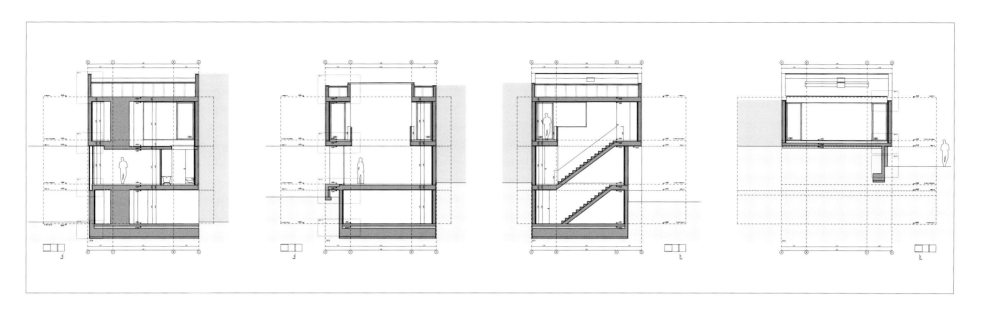

Fragmented House

Architect: AQSO Arquitectos Office / Pinés & Jové
Location: Laguna de Duero, Spain
Ground Floor Area: 200 m²

The concept solution for this house starts from a compact block transforming, after several divisions and shifts, into an external broken—down form, result of the arrangement of the interior spaces.

The house is located near Laguna de Duero, a town situated in the municipality of "Tierra de Pinares", in Valladolid. The building is orientated facing west and distributed in two levels: living room and day use spaces in the ground floor and master bedroom upstairs.

The form is conceived as a series of juxtaposed elements defining the different atmospheres and spaces. Therefore, the entrance is demarcated by two parallel blocks and another recessed one working as main access. The front part of the house, facing the garden, is marked out by the cantilevered block where the bedroom at the upper level is located, in contrast with the one of the ground floor. The rear of the house, where the garage can be found, is made up of several stair—shaped elements.

Inside, the living and dining rooms are linked into an open and continuous space just partially blocked by a stone masonry wall and the freestanding staircase giving access to the upper floor. From the master bedroom, provided with a generous walk—in wardrobe and ensuite Jacuzzi, it is possible to access the roof, partially used as terrace with a small solarium.

The facade is made by big scale matte ceramic pieces combined with stone masonry walls, inside there is a predominance of light colors in walls, floors, doors and windows.

In the front garden of the house there is a slender swimming pool with spa and an independent block facing the yard with a wide bay window to be opened and converted into a summerhouse.

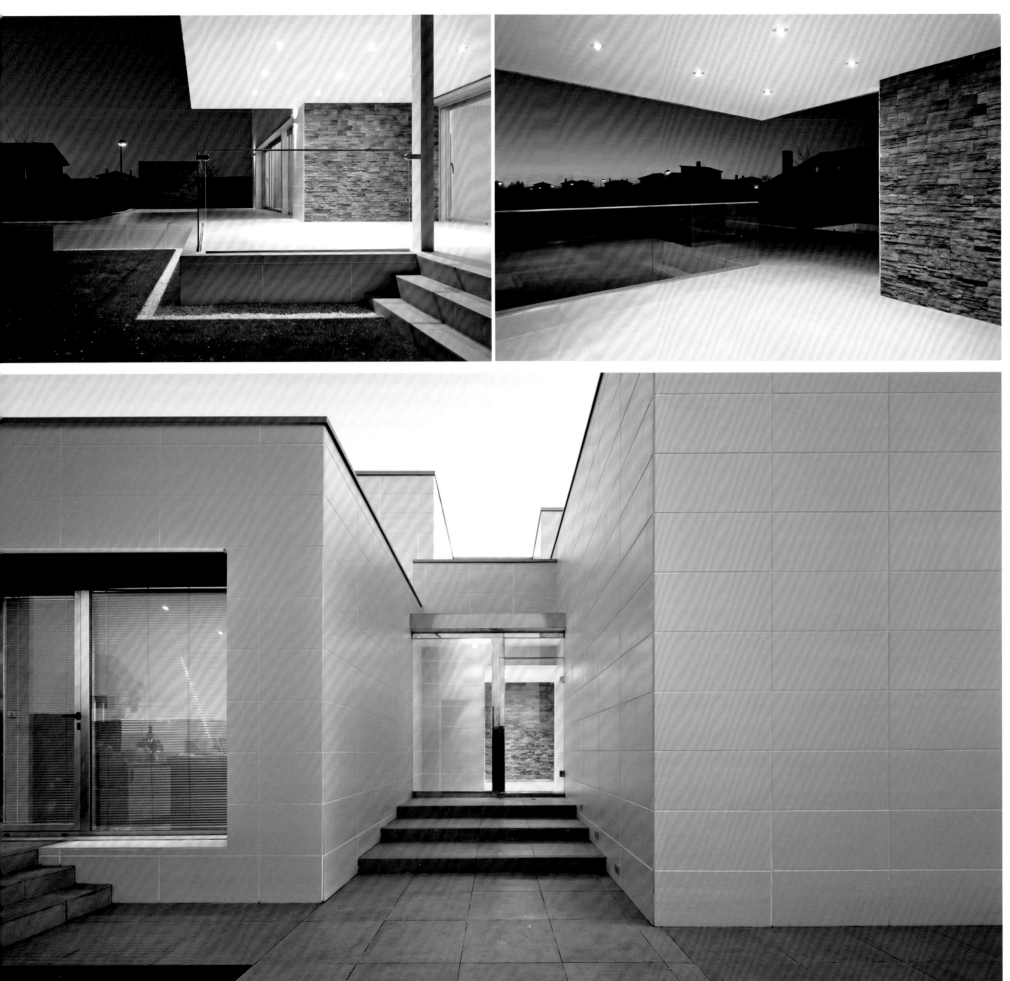

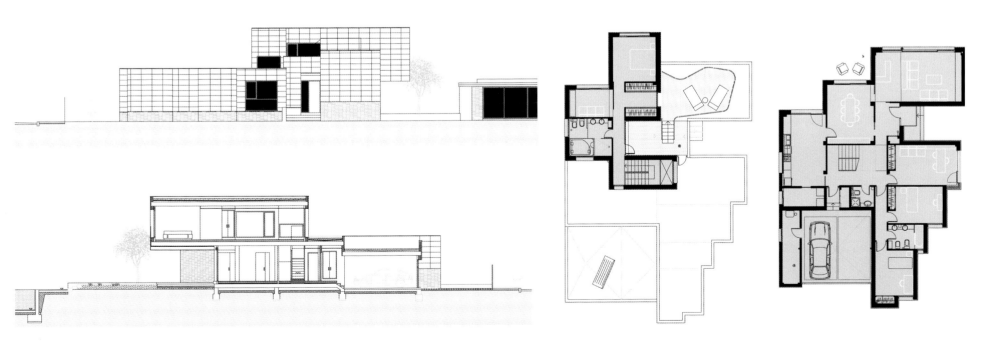

Can Bisa House

Architect: Batlle i Roig, arquitectes
Location: Barcelona, Spain
Photography: A. Flajszer

Can Bisa is a late—19th century mansion now owned by Vilassar de Mar Council. Situated on the Riera de Cabrils watercourse, it occupies part of a street block that used to include a factory, now demolished. Its historic and heritage value and the strategic position it occupies in the town as a whole led the Council to consider it the ideal venue for a cultural facility, completing the complex with a social housing block.

During the implementation of the facility's programme it rapidly became apparent that it would be impossible to accommodate all the necessary uses in Can Bisa, which is basically given over to council offices. the auditorium, bar and meeting rooms are therefore arranged on the ground floor of the residential building.

For the dwellings, the General Plan specified two volumes of different heights, joined to form an L—shape that respects the width of the streets and mounted on a larger base which is envisaged to accommodate the auditorium. The aim of the project was to unify the volumes by means of a single sloping roof, construct a continuous facade onto the streets, leading to the dwellings, and lay out the interior volume to encourage interrelation with the mansion. The passage generated between the two buildings becomes the main entrance to the complex. Its dimensions are more suggestive of a courtyard and, together with the gardens around Can Bisa, it comprises a series of leisure and recreation spaces that complement the activities offered by the facility. Ceramic planters designed specially by the artist Carme Balada serve to organize the complex.

Both the housing block and the mansion are given a textured finish of white stucco, in strips of varying widths. The balconies and shutters are made of metal, with masonry lattices. Part of the old factory is retained in one of the facades to bear witness to its past.

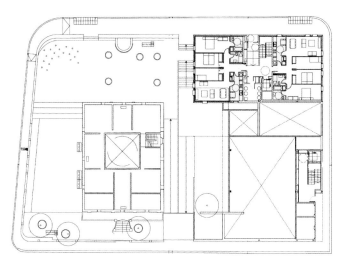

PLANTA BAIXA
GROUND FLOOR PLAN

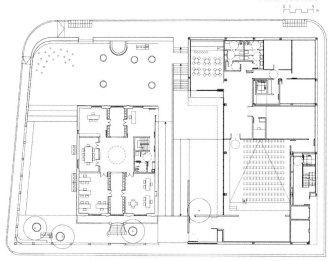

PLANTA SOTERRANI
UNDERGROUND FLOOR PLAN

Rehabilitation of Penthouse for Art Collectors

Architect: Clavel Arquitectos
Location: Murcia, Spain.
Total Area: 234,30 m²
Photography: David Frutos

Clavel Arquitectos is asked to refurbish an existing attic in the city. The main requirement from the client is to design an space for a living and for display his personal collection of paintings. On the other side, the architects want to evoke the feelings of a suburban house and take them into the city.

The refurbishment is made on an existing penthouse that is affected by strict regulations that imped to modify the alignment of the facade walls. So, as an evolution of the previous layout (very closed, with corridors and room on both sides), The architects proposed a system of habitable bubbles by covering and breaking the existing outer walls, that, thanks to the space that surround them, blur the boundaries between the inside and the outside. Each bubble hosts a different use: dining, living, main bedroom, children bedroom... while the space in between works as service and circulation.

The architects proposed a small number of materials, making sure that they are as durable as possible.

In this way, the bubble's walls are cladded with opaque white glass. Apparently contradictory with the creation of an art display system, this constructive solution gets very suitable when integrating three lines of flush—mounted rails for hanging paintings. The floor that surrounds the bubbles, that means the terrace as well as the circulation spaces, is paved with long slabs of grey phyllite stone, while their interiors are finished in different solid wood flooring.

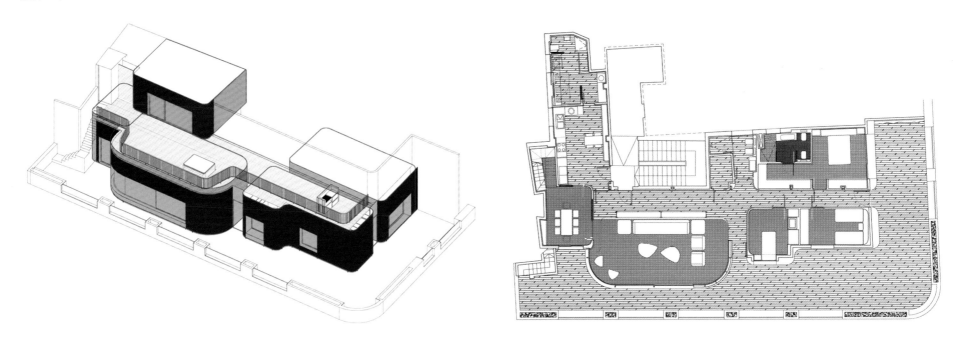

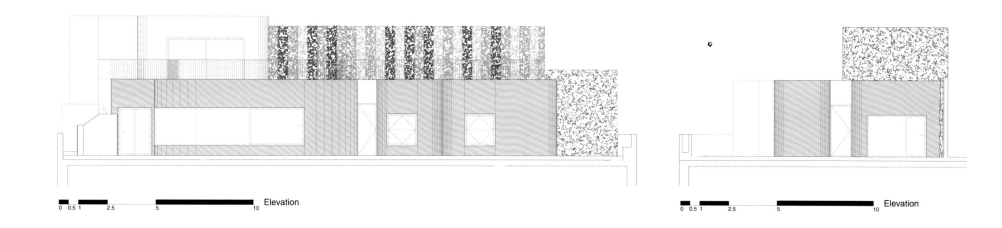

In the circulation spaces, these materials are complemented with a continuous perforated gypsum ceiling, in order to counteract the acoustic behavior of the glass.

Finally, the architects paid special attention to lighting, solving three different levels of use: daily use lighting, by downlights and recessed LED strips around the perimeter; art display lighting, by cardan spotlights; and signalizing lighting in the outside, by LED strips under the glass walls that make bubbles float at night.

Atrium House

Architect: Fran Silvestre Arquitectos
Location: Valencia, Spain
Site Area: 1,150 m²
Photography: Fernando Alda

A House in an urban area of the desire to maximize the feeling of spaciousness. Two strategies are used. The principal is to release the largest possible in the middle of the site allowing you to enjoy a private space with a height and volume incalculable. It enhances the perimeter of contact with the outside, land and housing understood as a continuum. On the other hand uses the existing slope to the ravine next to illuminate the basement, which enables you to host the program.

The building is developed along the southern and western boundaries of the parcel, which together with the elements of urbanization of the site, form a kind of atrium, with diagonal flight to a distant vision of the Sierra Calderona.

Access is accompanied by the south facade to find the point of intersection. At this point of view inside the distributor is located next to the stairs and the kitchen forms the backbone of the operation of housing. The southern zone where the rooms are available during the day, dematerialized their presence due to the overhead light. In the west the rooms fall to a portion of parcel with a more domestic scale, while the master bedroom overlooks the lift light reflected on water.

The basement and garage are in the dark cellar. All other uses of the program look into the ravine through which light up.

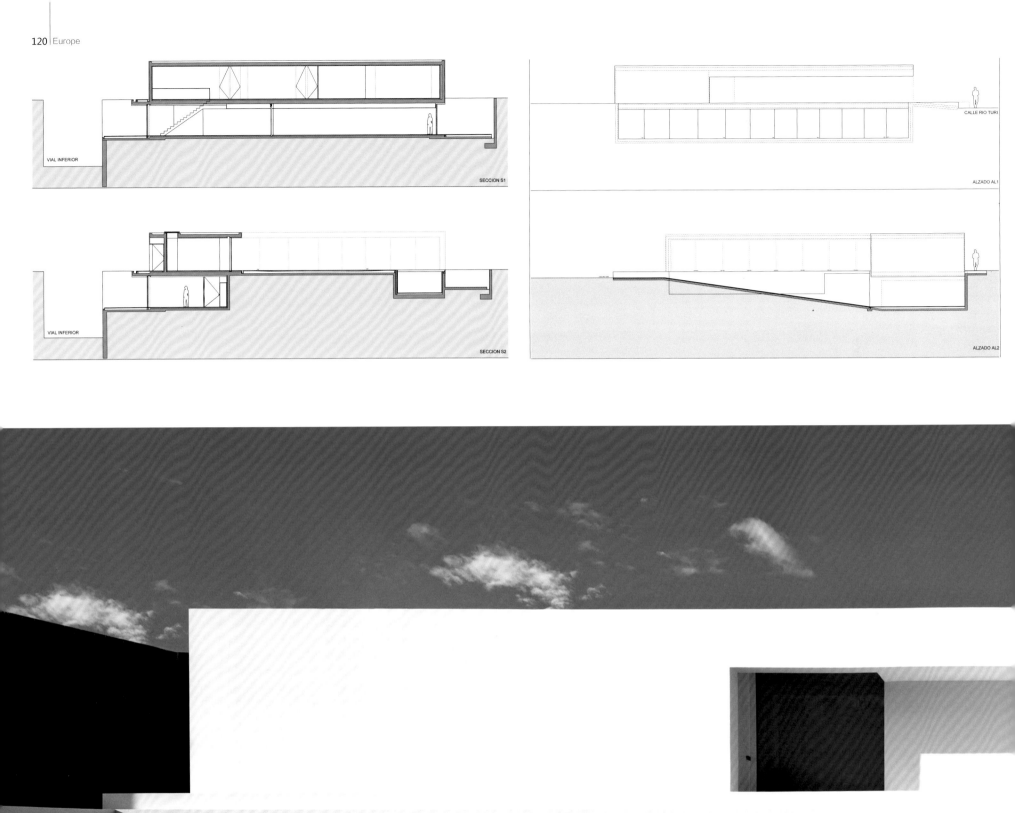

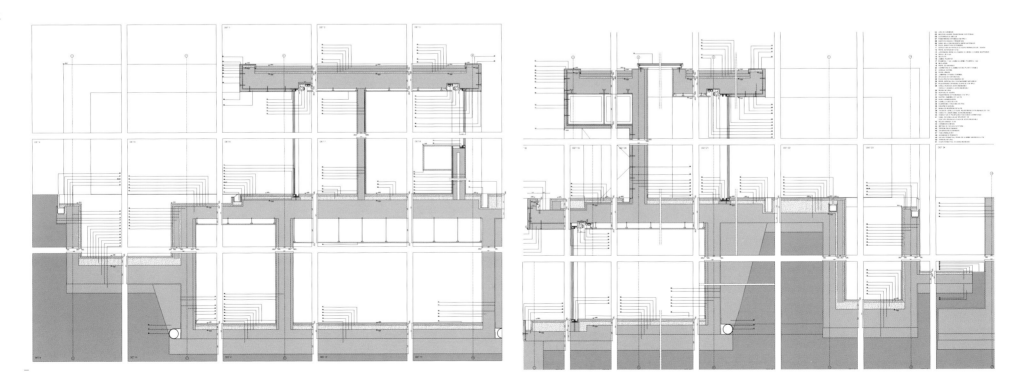

Dwelling in Etura

Architect: Roberto Ercilla
Location: Etura, Alava, Spain
Area: 218 m²

The parcel slope determines the starting point of the project. The visual impact of the housing is reduced by placing it below the access level into the side, with a piece strongly over hanged.

This is achieved by minimal intervention in the environment (9% parcel occupancy). The roof garden with a small access pavilion and vehicle protection completes the intervention.

The buried provision, vegetation cover, the use of renewable energy—biomass—and water—saving measures, greatly reduces energy consumption.

The shot passed through is resolved by the ladder, which acts as a chimney.

The house is oriented south, coinciding with the sights and disposition prevents the North and the cold winds of winter. All construction is reinforced concrete structure.

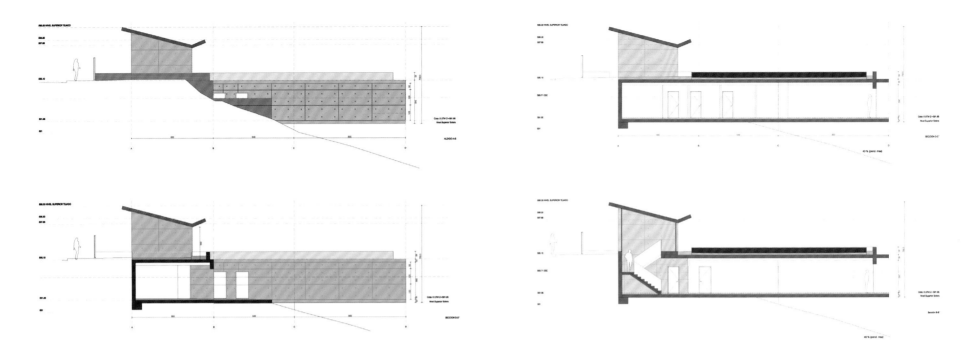

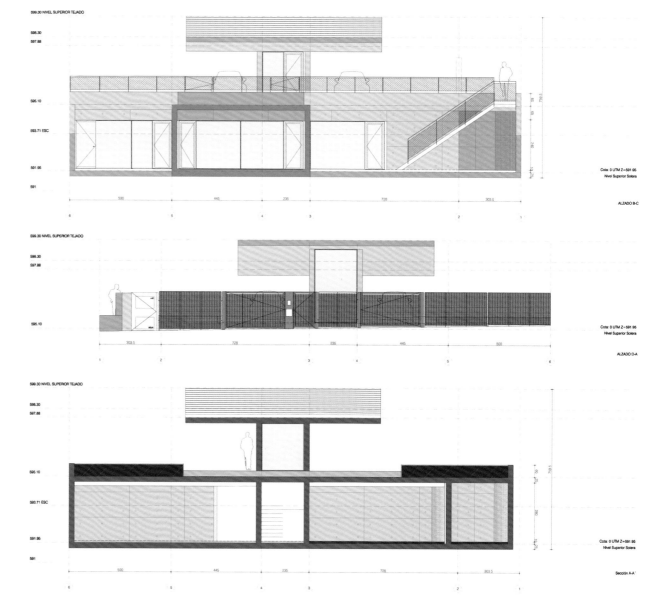

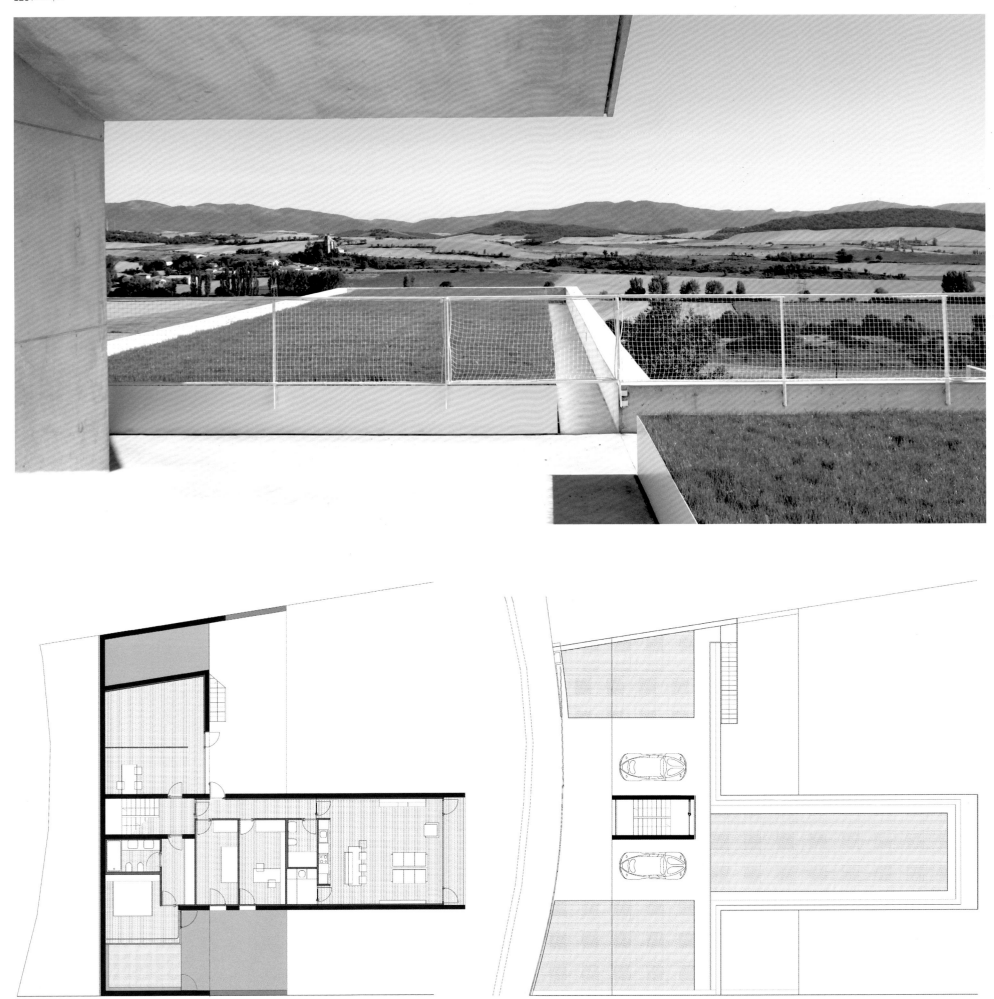

Villa BH

Architect: WHIM architecture
Location: Burgh—Haamstede, The Netherlands
Site Area : 1,751 m²
Photography: Sylvia Alonso

Villa BH is a modern, (environment) friendly house with a remarkable experience of space, light and the natural context.

The villa is positioned on a rectangular plot of 34.75m x 50m, that is enclosed at 3 sides with similar plots and freestanding houses. On the back (north east) of the plot there's an old embankment with several tall trees, whose existence is protected by local regulations. From the living program; the kitchen, dining area and living are all orientated on this embankment with the large trees. Here the villa has a facade width of 20 meters.

Villa BH is inhabited by a couple 60+ of age. To optimize the accessibility of the house all the program is situated on the ground floor level around a patio. This enclosed outdoor space provides the owners of the house the privacy they admired. As at the same time the patio makes the living area an enlightened space and gives it a facade to the south.

On the other side of the patio is situated the main bedroom. By making the facade of the patio totally from glass panels, the main bedroom has a great see through towards the existing embankment with the several tall trees as a central focus point on the plot.

The ceiling of the living area has an extra height in the shape of a sloped roof. The physical appearance of this area becomes hereby more specific and highly qualitative. Lifting the roof in this area also allows perspectives to the existing treetops, that give this plot its specific character, from all the different areas inside the building.

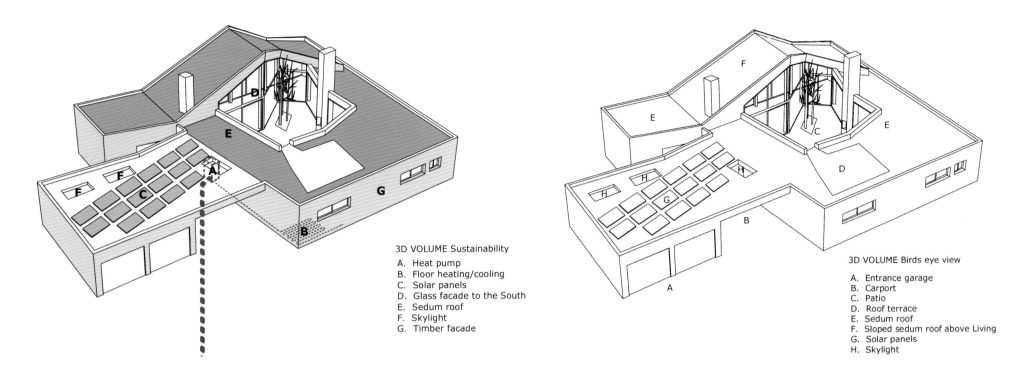

3D VOLUME Sustainability

A. Heat pump
B. Floor heating/cooling
C. Solar panels
D. Glass facade to the South
E. Sedum roof
F. Skylight
G. Timber facade

3D VOLUME Birds eye view

A. Entrance garage
B. Carport
C. Patio
D. Roof terrace
E. Sedum roof
F. Sloped sedum roof above Living
G. Solar panels
H. Skylight

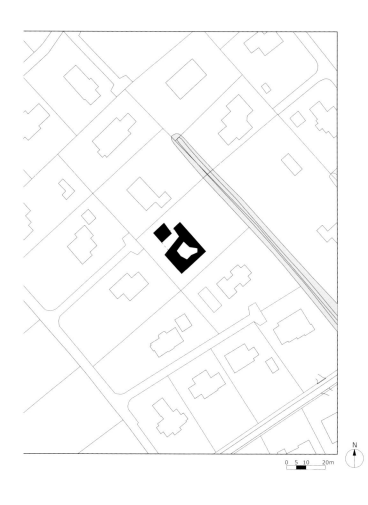

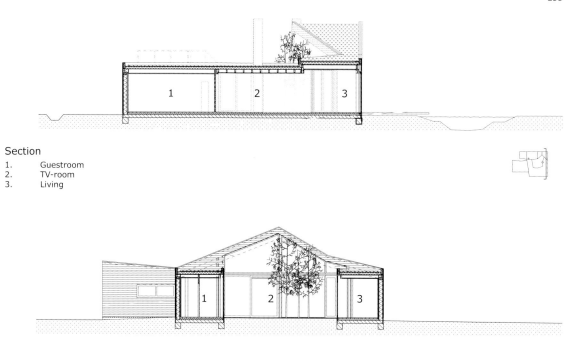

Section
1. Guestroom
2. TV-room
3. Living

Section
1. Corridor
2. Patio
3. TV-room

The villa is designed as environment friendly with extra insulated facades, with 30cm of insulation. With this thick insulation there's a timber construction, that suits the thickness of the package. The roof is extra insulated and covered with sedum, which also regulates the distribution of the rainwater more gently. On the flat roof are 20 solar panels for electricity. A heat pump warms the interiors in the winter and cools them in the summer with natural temperature differences retrieved deep in the ground. As an extra heating there are 2 fireplaces for wood, one in the living and one in the TV room.

Villa BH is an environment friendly design that emphasizes the natural qualities of its context. The design was made over 2008—2009 by Ramon Knoester. At the end of 2010 his garden design was also completely realized.

V12K0709 Piano House

Architect: Pasel Kuenzel Architects
Location: EWR —terrein, Leiden, The Netherlands
Area: 259 m^2 (excl. patio, terraces and garden)
Photography: Marcel van der Burg, primabeeld

This urban villa for a pianist and a violinist, is located on the site of a former industrial area, in the heart of the Dutch city of Leiden.

The spatial idea of this urban residence is based on a 3 metres high, all—embracing wooden screen that surrounds the whole site enclosing as well the building volumes as the building voids of the patio and garden.

The composition of this wooden filter, made out of dancing' timber fins, refers directly to the musical oeuvre of the concert violist and the pianist living in the house. It manifests multifaceted visual relationships and allows for an exceptional relation between public and private life, between inside and outside of the house.

Named V12K0709, the house occupies a rectangular plot and screens two courtyard gardens behind its slatted timber exterior. Most of the ground floor spaces open out onto these gardens, including a music room where the occupants can practice their instruments. Two more terraces are located on the first floor and overlook those below from opposite corners.

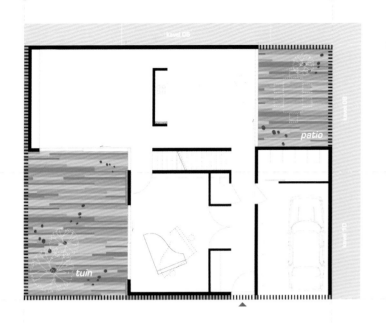
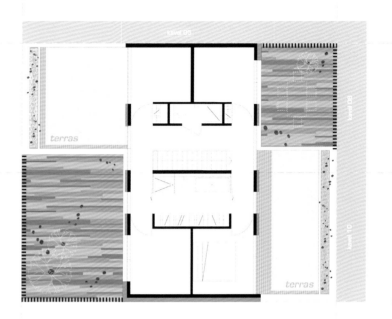

House V12K0102

Architect: Pasel. Kuenzel Architects
Location: Leiden, The Netherlands
Area: 240 m²
Photography: Marcel van der Burg

On the site of a former slaughterhouse in the historical heart of the Dutch university city of Leiden, emerges one of the biggest urban developments of private dwellings in the Netherlands. In their series of eleven, the Rotterdam based architects, Pasel.Künzel Architects, present yet another spectacular house, giving a new interpretation of the classical Dutch housing typology.

With their V12K0102 residence Pasel.Kuenzel Architects created a remarkable project on an almost triangular building plot, the remnant of an inner city housing block. On a 30 meters long one—story high base, two building volumes are placed on opposite sides, one being the "children's house" and the other serving as the "Parents house". The two parts facing each other allow a connection with visible eye contact, but are physically separated.

Collective spaces for living, dining and playing are situated on the ground floor, meandering around two intimate courtyards. This setup establishes an immediate relation between "life inside and outside", creating an oasis in the city. Towards the city, the introverted house reveals its inner side with two gigantic glass panes that also permit Dutch light to reach deep into the museum like spaces.

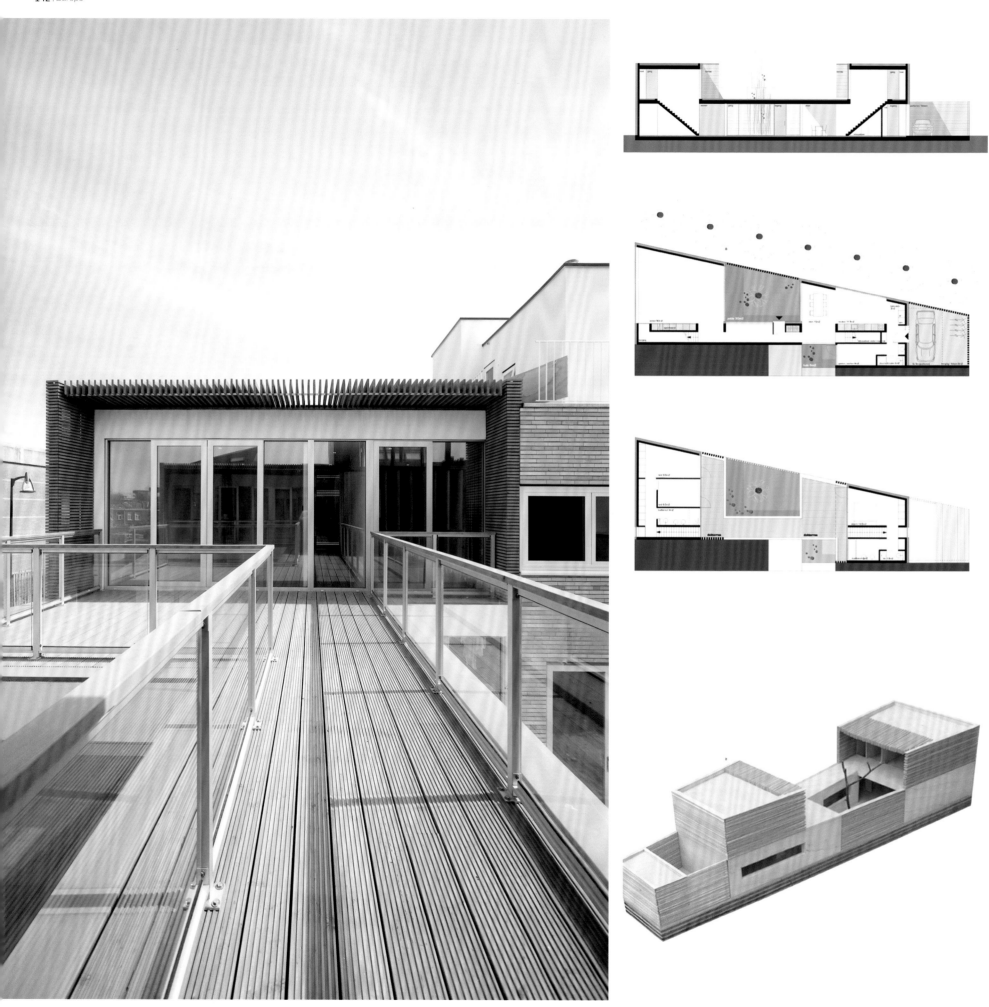

Electric Boathouse

Architect: Sebastiaan Jansen Architectuur
Location: Heeg, Friesland, The Netherlands
Floor Area: Net floor area approximately 695 m²
Photography: Sebastiaan Jansen

From the Friesian perspective a remarkable building has arisen in Heeg. The boathouse is not only striking in its appearance, the building is also exceptionally sustainable. In this design, context, sustainability and architecture meet in an extraordinary symbiosis, for example solar panels are integrated into the roof of the main volume. The electric boathouse is the first building in the Netherlands where electric sailing and architecture are brought together.

"We call it the building blocks," says the neighbour, while a passing driver stops and looks up with wonder. The neighbour continues, "it is completely different from the usual commercial units that you see here. It takes some time getting used to, but it is striking".

The building accommodates a warehouse with a workshop for the building and rebuilding of electric boats, an exhibition space where the boats can be exhibited whilst floating and an office. These three functions have their own appearance in form as well as style. Through the arrangement of the three primary functions around a central circulation axis, they are well connected, with only short walking distances between. At the front of the building the transparent circulation axis extends outside the building where it operates as a visitor entrance and exhibition space.

The volume that appears to float above the completely glazed exhibition space has come into being through a play on context and sustainability. The skewed plot is legible in the floor surface, whilst the upper surface orientates itself towards the sun. This creates a sculptural, architectural gesture, with a kink in the front elevation. The building reaches it's climax in this corner, where the black, heavy volume accentuates the corner of the plot. By extending this outermost corner, it protrudes over the building volume like the bow of a ship.

By sloping the roof surface towards the south, an optimal surface is created for the application of photovoltaic cells. 72 solar panels have been placed on the corner volume (approximately 91m^2) delivering approximately 11,072 kWh per year, enough to recharge an electric boat 1,385 times. In addition to the solar panels, a number of sustainable applications have been incorporated into the building; rainwater is collected in a gray water tank and is used in washing the boats and flushing the toilets. The heating and cooling of the building is through an under—floor heating system, linked to a geothermal heat pump.

The "cradle to cradle" principle is carried through in the construction and materialization of the building. Where possible, the assembly is with mechanical connections, so that very little energy is required in disassembly. Furthermore, materials have been used in their purest form. Only a few materials are finished with paintwork, instead timber remains timber and stone remains stone.

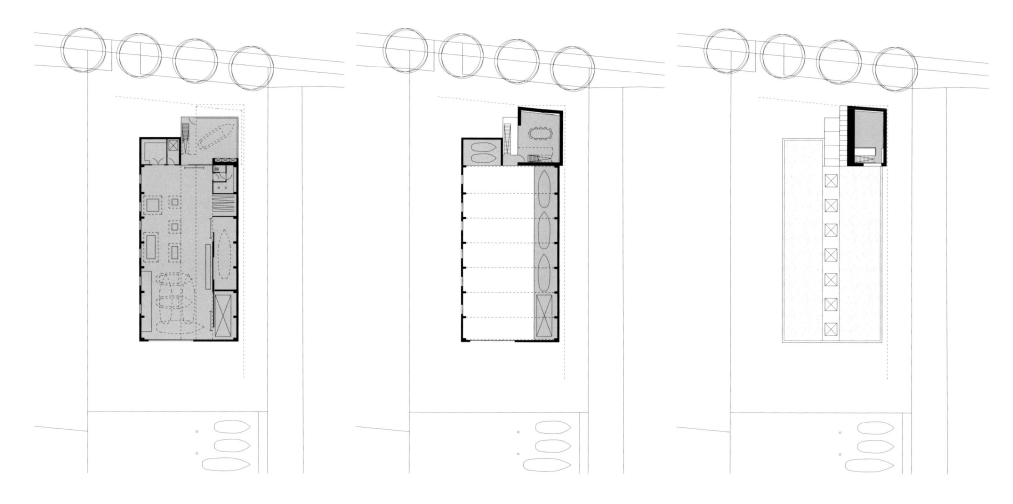

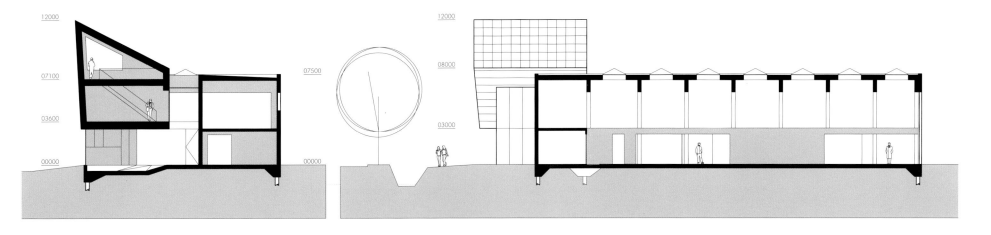

Water Villa

Architect: Framework Architecten & Studio Prototype
Location: Amsterdam, The Netherlands
Area: 250 m²
Photography: Jeroen Musch

This water villa is designed by Framework Architecten & Studio Prototype for a waterfront location near the Olympic Stadium in Amsterdam.

The relation between the water and house is central to the design. There is a subtle playfulness between open and closed. The vertically designed pattern, an abstract allusion to the water, provides not only optimal privacy but also a subtle play of light inside the residence itself. The inhabitants are able to regulate their privacy by, for example, an integrated folding window that can be opened and closed by remote control.

The house is spacious with three levels, one of which is below the water, while living and work areas are located above the water. The three levels are spaciously connected by an inner patio, which not only centrally organizes the plan of the house but creates sufficient light in the lower level as well. Also, the steel staircase that has such distinctive significance for the character of the house, is located in the patio. Here again, the vertical pattern of the staircase, consisting of a steel stripe pattern, provides a dynamic display of light and direction.

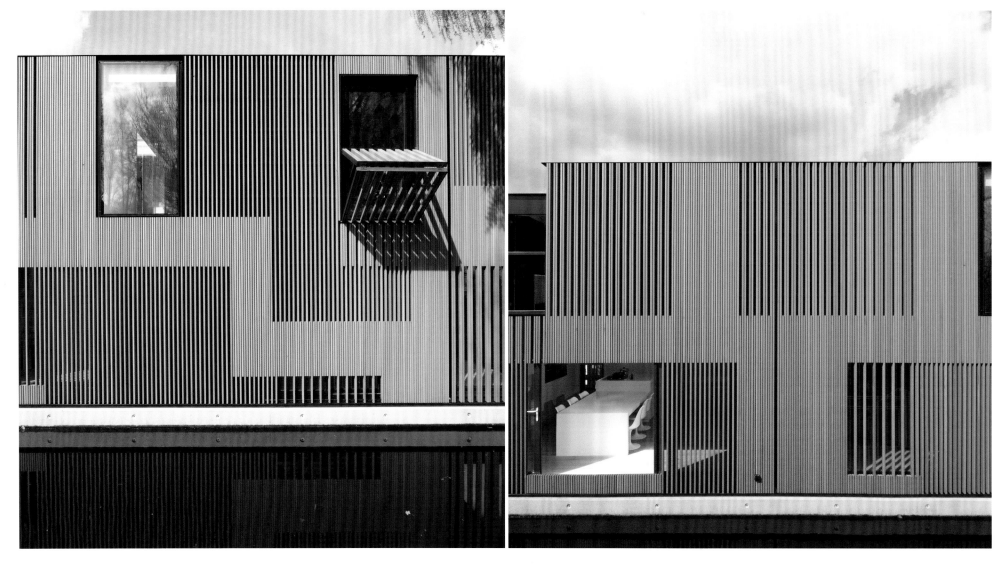

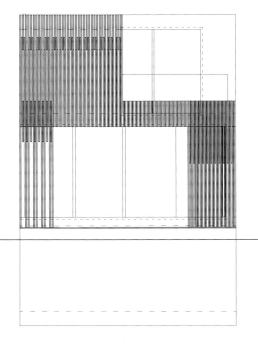
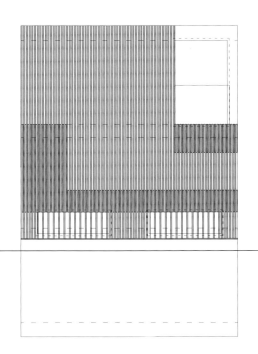

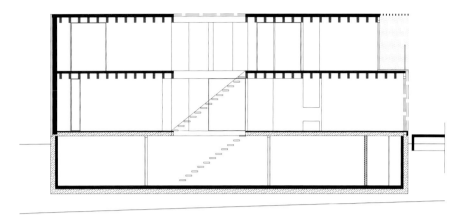

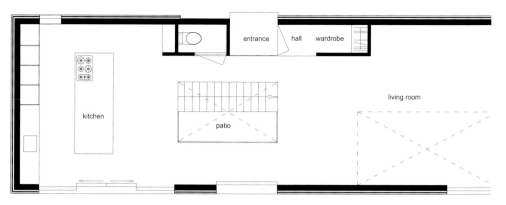

GROUND LEVEL

FIRST FLOOR

BASEMENT

The Balancing Barn

Architect: MVRDV
Location: Thorington, Suffolk, UK
Area: 210 m²
Photography: Edmund Sumner, Courtesy of MVRDV
and Living Architecture

Balancing Barn is situated on a beautiful site by a small lake in the English countryside near Thorington in Suffolk. The Barn responds through its architecture and engineering to the site condition and natural setting. The traditional barn shape and reflective metal sheeting take their references from the local building vernacular. In this sense the Balancing Barn aims to live up to its educational goal in re-evaluating the countryside and making modern architecture accessible. Additionally, it is both a restful and exciting holiday home. Furnished to a high standard of comfort and elegance, set in a quintessentially English landscape, it engages its temporary inhabitants in an experience.

Approaching along the 300 meters driveway, Balancing Barn looks like a small, two-person house. It is only when visitors reach the end of the track that they suddenly experience the full length of the volume and the cantilever. The Barn is 30 meters long, with a 15 meters cantilever over a slope, plunging the house headlong into nature. The reason for this spectacular setting is the linear experience of nature. As the site slopes, and the landscape slopes with it, the visitor experiences nature first at ground level and ultimately at tree height. The linear structure provides the stage for a changing outdoor experience.

At the midpoint the Barn starts to cantilever over the descending slope, a balancing act made possible by the rigid structure of the building, resulting in 50% of the barn being in free space. The structure balances on a central concrete core, with the section that sits on the ground constructed from heavier materials than the cantilevered section. The long sides of the structure are well concealed by trees, offering privacy inside and around the Barn.

The exterior is covered in reflective metal sheeting, which, like the pitched roof, takes its references from the local building vernacular and reflects the surrounding nature and changing seasons.

On entering the Barn, one steps into a kitchen and a large dining room. A series of four double bedrooms follows, each with separate bathroom and toilet. In the very centre of the barn the bedroom sequence is interrupted by a hidden staircase providing access to the garden beneath. In the far, cantilevered end of the barn, there is a large

EAST FACADE

LONG SECTION

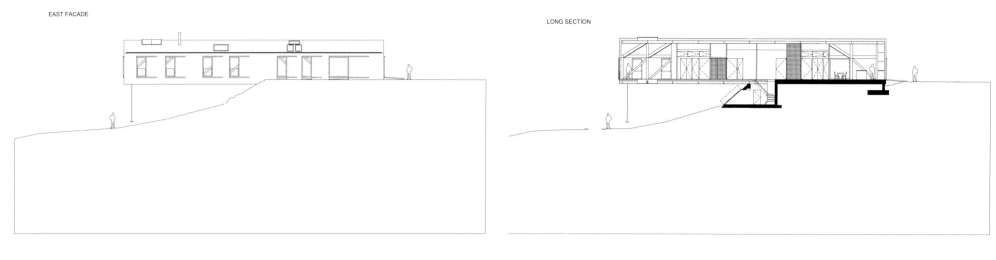

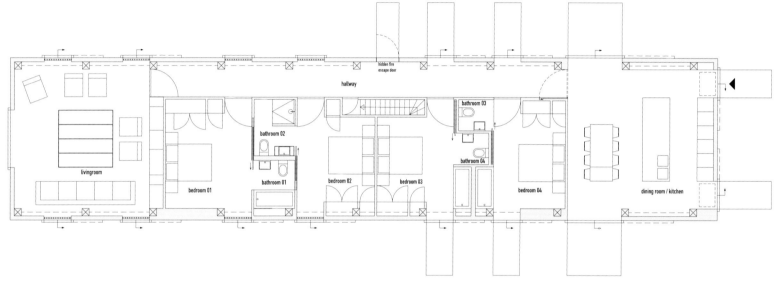

living space with windows in three of its walls, floor and ceiling. The addition of a fireplace makes it possible to experience all four elements on a rainy day. Full height sliding windows and roof lights throughout the house ensure continuous views of, access to and connectivity with nature.

The interior is based on two main objectives:

— The house is an archetypical two — person home, expanded in shape and content so that it can equally comfortably accommodate eight. Two will not feel lost in the space, and a group of eight will not feel too cramped.

— A neutral, timeless timber is the backdrop for the interior, in which Studio Makkink & Bey have created a range of furnishings that reflect the design concept of the Barn.

The rooms are themed. Partly pixilated and enlarged cloud studies by John Constable and country scenes by Thomas Gainsborough are used as connecting elements between the past and contemporary Britain, as carpets, wall papers and mounted textile wall — elements. The crockery is made up of a set of English classics for two, and a modern series for a further six guests, making an endless series of combinations possible and adding the character of a private residence to the home.

The Barn is highly insulated, ventilated by a heat recovery system, warmed by a ground source heat pump, resulting in a high energy efficient building.

House at the Edge of a Forest

Architect: Hilberink Bosch architects
Location: Heesch, The Netherlands
Land Area: 1,600 m²
Photography: René de Wit, Paul Kozlowski

The house, situated on a beautiful lot at the edge of the forest, consists of two different volumes: an L—shaped base on which an oblong volume balances. Together they form a sculpture which resembles a fallen tree on a pile of earth.

The public functions of the house are situated in the L—shaped base. The outside walls of the L—shape which face the public road look unapproachable and secretive. The wall is made with long, dark, robust bricks emphasizing the horizontal lines.

The interior of the house is open and light. The living space is connected with the terrace, the garden and the forest and a flood of light is entering the house. The garden facade of the house is formed by a concrete structure, the imagination of modern living within the rampart.

On this basement a timber volume is placed in which the more private rooms such as bedrooms and bathrooms are situated. The wooden volume resembles a fallen tree, balancing on the firm base. The steel structure of this volume has been clad with wood out of Louro Preto, an FSC certified tropical hardwood.

The wooden volume protrudes far beyond the base, forming sheltered places around the house. On one side the timber volume is firmly anchored to the ground with a glass volume. Angled and sturdy steel columns protect the glass. On the garden side, the wooden volume forms a seven meters wide overhang. This overhang provides shade and frames the terrace forming a continuum of the interior, a space between inside and outside.

All the edges of the different volumes are made without any eaves, the material dissolves in the air. This reinforces the abstract appearance of the sculpture.

Just as a wanderer, caught in a thunderstorm, and seeking shelter under a fallen tree, the inhabitants will find protection in this house.

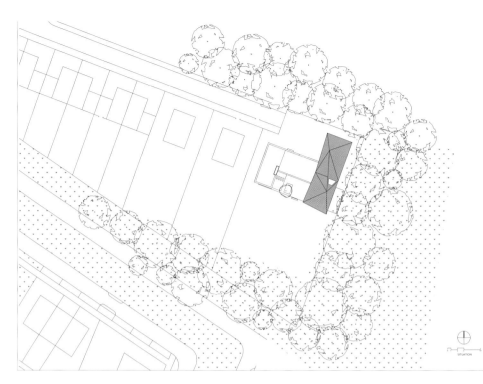

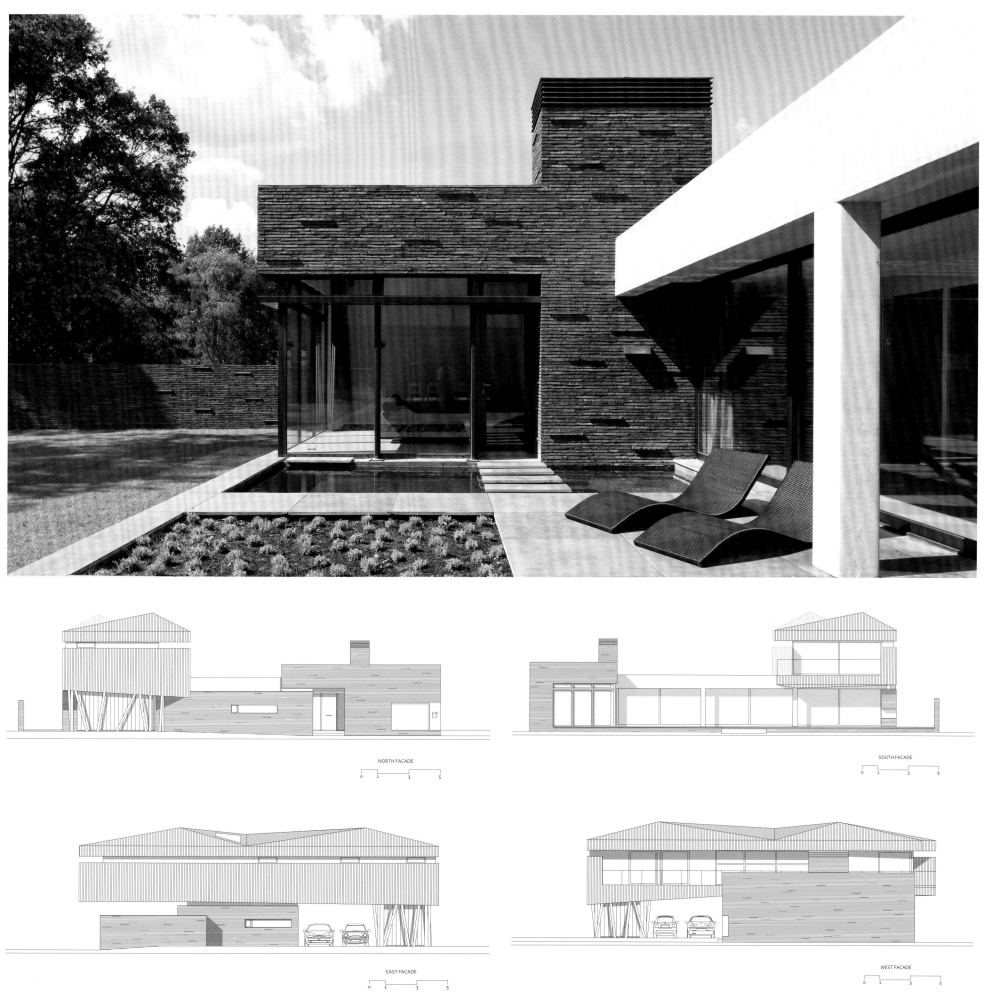

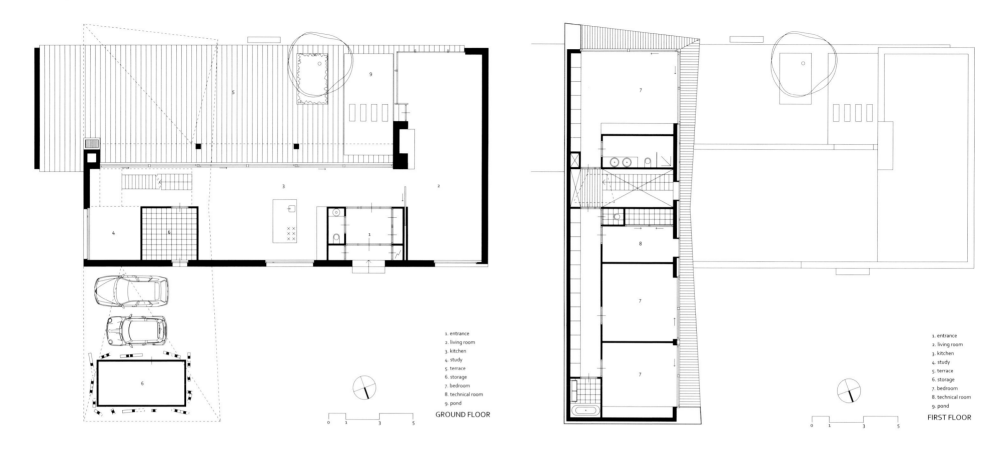

1. entrance
2. living room
3. kitchen
4. study
5. terrace
6. storage
7. bedroom
8. technical room
9. pond

GROUND FLOOR

1. entrance
2. living room
3. kitchen
4. study
5. terrace
6. storage
7. bedroom
8. technical room
9. pond

FIRST FLOOR

Villa Wageningen

Architect: Mecanoo Architecten
Location: The Netherlands
Area: 724 m²
Photography: Christian Richters

It is a three—level residential villa with one level freestanding garage and studio building, total 724 m². This private villa is beautifully situated on a hill in the natural landscape region of the Veluwe. The residents enjoy magnificent views of the Rhine River from their living room. The design for the villa takes its inspiration from the beautiful views from the living room and main bedrooms and also from the spacious kitchen/lounge area toward the big garden. Large terraces elevate the house to just above grade, making the house appear as if it is floating. The villa's transition from interior to exterior is gradual and a canopy surrounding the whole house provides beautifully framed views while allowing the outdoor spaces to be enjoyed in the summer and fall. The round forms of the villa express the wish of the family to stay close to nature while simultaneously realizing an individual design.

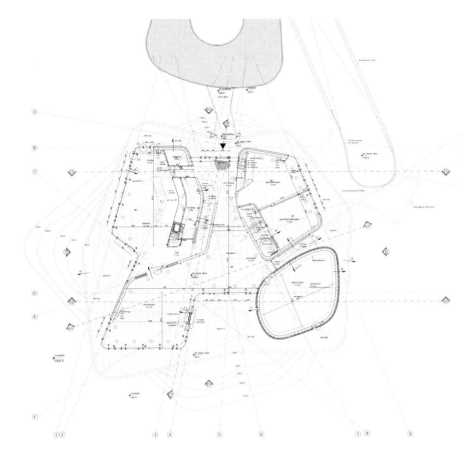

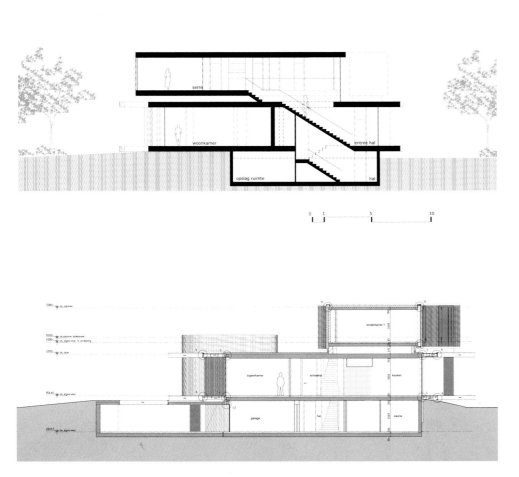

Villa Frenay

Architect: 70F architecture
Location: Lelystad, The Netherlands
Photography: Luuk Kramer

The project consisted of a detachs house with outbuildings on a plot in villa park "de Noordzoom" in Lelystad, at the foot of the Enkhuizen—Lelystad dike. The client wants, like many others, light, air and space.

The house is a one—story bungalow, while the site asked for a response to "living at the water". The latter is accomplished by making a veranda along the entire length of the building (south), at the most important rooms of the house. from bedroom, to wardrobe, to the master bathroom, to hobby room, to the kitchen and finally to the dinning—and around the corner the living room. Along this line, there is a transition from water to veranda, to completely glass facade, with doors behind which all mentioned areas are.

To break the flat polder landscape, the roof of the living/dining room is slightly sloping. At the non—waterfront (north) of the house there are the children's rooms with their bathroom, a utility room and the living room. The existing artificial "hills" on the premises where the architects accentuated, and the carport with storage room is half buried in one of them. The sauna with garden shed is placed on another hill, and is the highest of the three buildings. From the sauna is a view over a Cor—Ten steel waterfall that runs along the open—air terrace at the gable of the house, slopping towards the water feature that runs along the plot.

With the spread of the buildings, a residential area is created, more than a house with outbuildings. The transparency and direction can be seen in the interior, providing the ultimate inside—outside experience, emphasized by the slopes on the plot. The structural challenge of the T—shaped gable wall with glass corners, results in a phenomenal experience of the indoor and outdoor space in the living and dining room. The structural and architectural tension of the solution in this wall is experienced almost everywhere in and around the house.

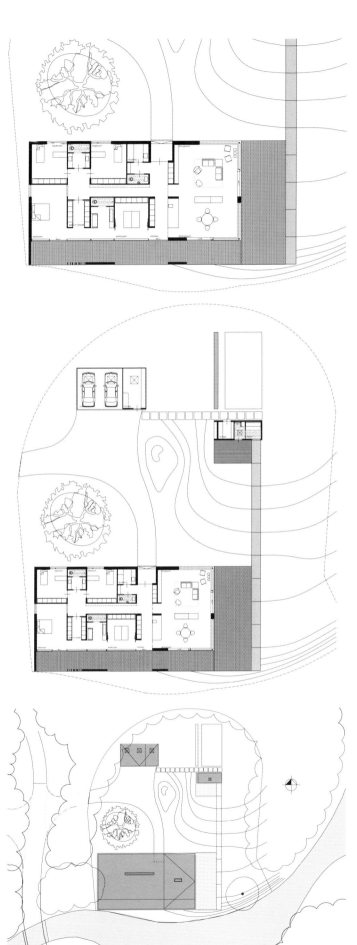

Villa Biesvaren

Architect: BBVH Architecten BV
Location: Den Haag, The Netherlands
Site Area: 500 m²
Photography: Luuk Kramer

This modern villa, design by Rotterdam bases architecture studio BBVH Architecten, is located in a new suburb of Den Haag and designed for the owner of a furniture company. The dynamic architecture is characterized by its dark colors and large cantilevering terraces. Oriented at the waterfront, the house is designed to enjoy the view of the surroundings.

The house is built with a steel structure around a concrete core, and filled with lightweight prefab timber exterior walls. This allows the terraces to cantilever 3 meters. Used materials are gray glass fibre reinforced concrete panels in combination with an anthracite plaster. The ceiling under the cantilevering terraces is finished with a warm brown wooden cladding.

Contrasting the dark exterior is a bright white interior with spacious rooms. The ground floor has a modern open kitchen and a large living room overlooking the water. Three children's bedrooms and a bathroom are located on the first floor, with access to a large terrace. The master bedroom with bathroom and walk—in closet is located at the top floor, connected to a private terrace to enjoy the evening sun.

gevel OOST

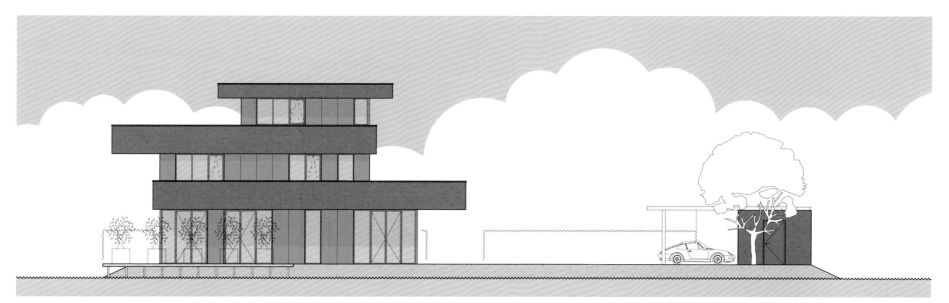
gevel WEST

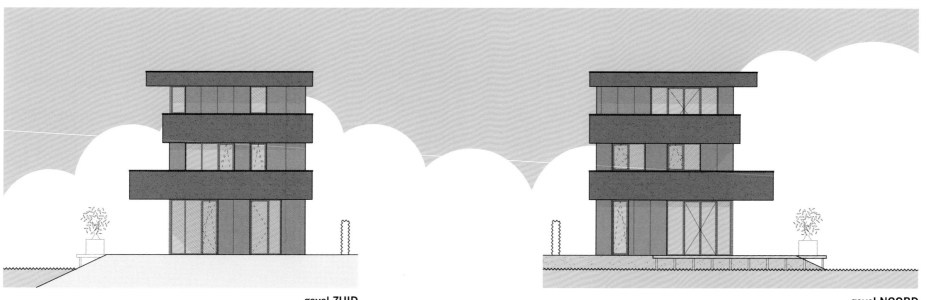

gevel ZUID

gevel NOORD

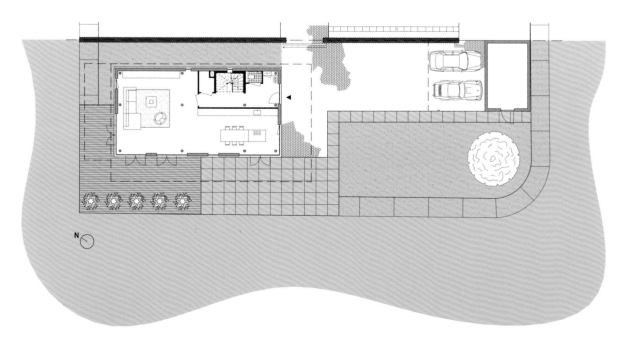

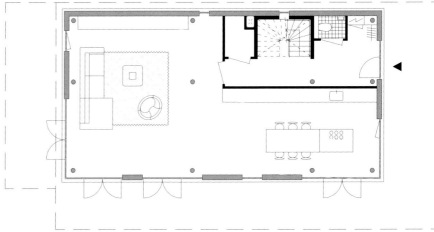

nivo 00

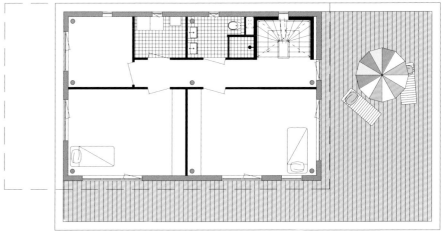

nivo 01

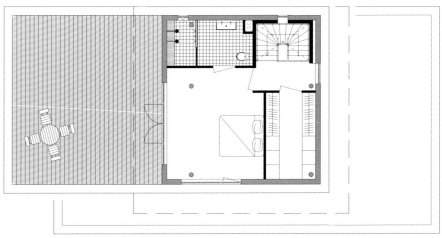

nivo 02

Villa Festen

Architect: BBVH Architecten BV
Location: Den Haag, The Netherlands
Site Area: 470 m²
Photography: Luuk Kramer

This modern villa, designed by BBVH Architecten, is located in the Biesland area, a new suburb of Den Haag. The area is characterized by traditional brick architecture but the urban plan prescribed a few private housing developments at specific points in the area. This private house is located at the entrance road in the north east side.

The organization of the house and its position on the plot is aimed at the sun. The entrance, garage, bathrooms and storage rooms are all located at the north east side. The living room, dining room, bedrooms and the terraces are all located at the south side. A spacious garden on the sunny side is created by moving the house as much as possible to the north side of the plot. The living room has a glass facade with large sliding doors.

On the ground floor, the north corner has been saved in order to create a covered entrance and parking space. Also located on the ground floor are the living room, dining room and kitchen. The first floor has 4 bedrooms, 2 bathrooms, storage space and a terrace. The top floor has a studio and a large terrace overlooking the surroundings. A spiral staircase in a spacious vide connects the different floors.

The house has a dynamic volume by its terraces and cantilevering parts with a different character on all sides. The plinth and top floor are finished with white plaster. The volume of the first floor is cladded with a silver blue slate stone. The ceiling underneath the cantilevering first floor and some parts between the windows is finished with oiled western red cedar.

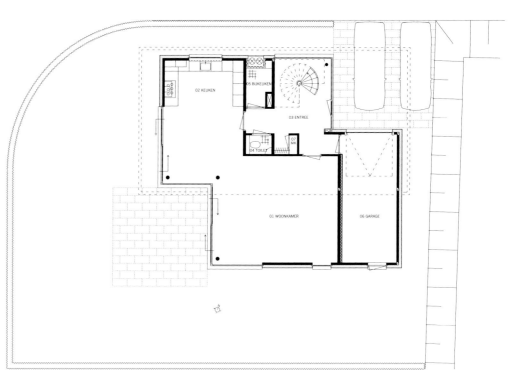

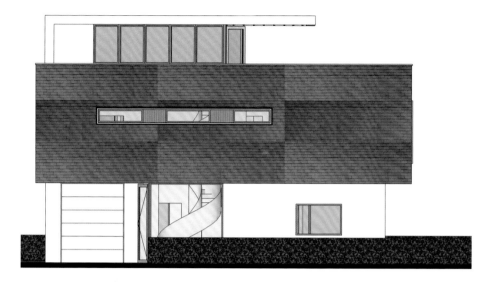
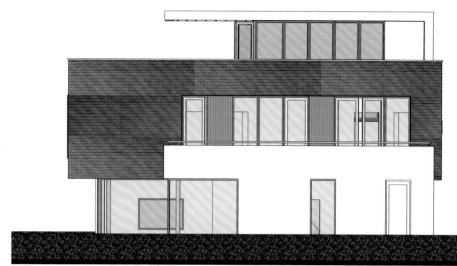
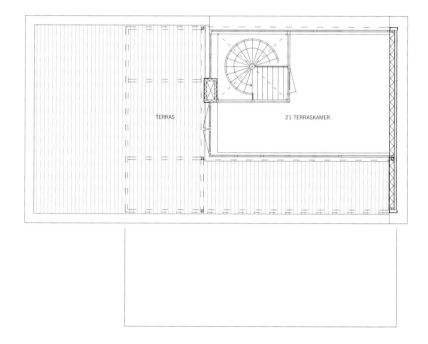

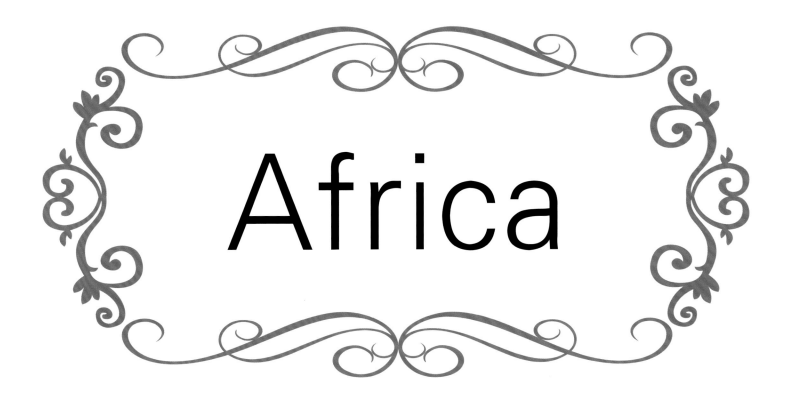

SGNW House

Architect: Metropole Architects
Location: Zimbali, South Africa
Photography: Grant Pitcher

The clients requested that the architects create a dream house for their site in the Zimbali forest estate. Several bodies of water, including Koi ponds, water features and a rim flow swimming pool appear to coalesce into one, and flow through the house and out into the forest. The stacked roof is fragmented and linked with flat roof slabs, in correspondence to the spatial arrangement of the rooms below, which both scales and articulates the massing of the house. The main bedroom suite cantilevers six metres over the patio below, providing protection from the weather, as well as "wow" factor. Large amounts of glazing optimize views of the indigenous bush that encapsulates the house, and together with the palette of raw materials including natural timber, off shutter concrete, water and natural stone cladding, offset the clean architectural lines with a warmth and Zen like ambiance.

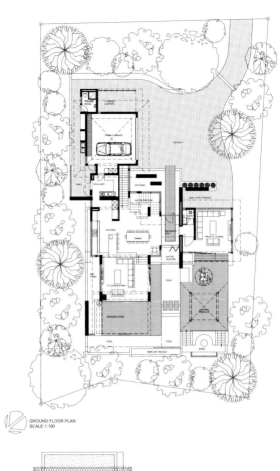

GROUND FLOOR PLAN
SCALE 1:100

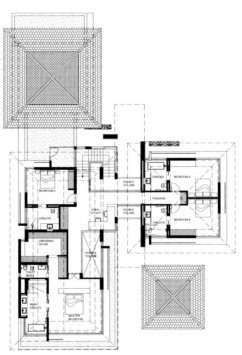

FIRST FLOOR PLAN
SCALE 1:100

PUMP ROOM PLAN
SCALE 1:100

RUMPUS ROOM PLAN
SCALE 1:100

Glass house

Architect: Nico Van Der Meulen Architects
Location: Johannesburg, South Africa
Site Area: 4,000 m²
Photography: Barry Goldman, David Ross

The house is situated on a 4,000 m² site, with a total floor area of 2,500 m². An existing house was demolished to construct the new house. The owner requested a modern, glamorous, open plan, light—filled house with views from all rooms into the garden.

The shape on the south side is a half circle, forming a horseshoe on the north side. Approaching the house from the gate the driveway is elevated to allow glimpses through the house to the garden and raised water feature on the other side of the house.

The porte cochere is a suspended glass and stainless steel structure, with view into the house and a stainless steel and glass staircase, suspended over a heated pond, (which in summer acts as a temperature stabilizer, and in winter as a giant heater) with a circular, raised glass water feature in the background, framed by a beam two storeys high.

To the right is a small sunken formal lounge, and to the left a timber—clad lift tower.

The dining room is raised a couple of steps above the family room. The window to the dining room is a 6m high curved glass enclosure, where each sheet of glass leans over further than the previous sheet, with glass fins holding it in position. The frameless glass folding doors starts at the dining room, and stretches for nearly 70m around the dining room, atrium, family room, lanai, indoor pool and gym.

The family room is partially double volume, flowing seamlessly into the lanai and heated indoor pool, with a bar, pizza oven, gas and wood braai. The kitchen leads off the family room and dining room, with a pair of automatic, frameless sandblasted doors leading from the dining room to the kitchen. A breakfast area and

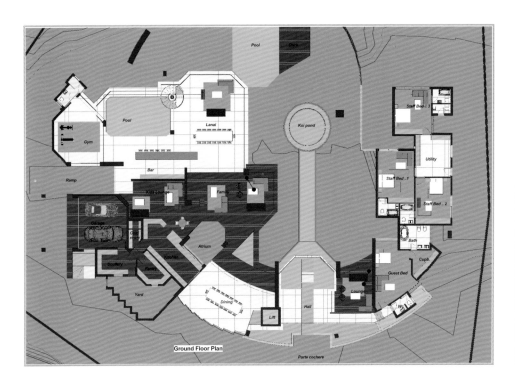
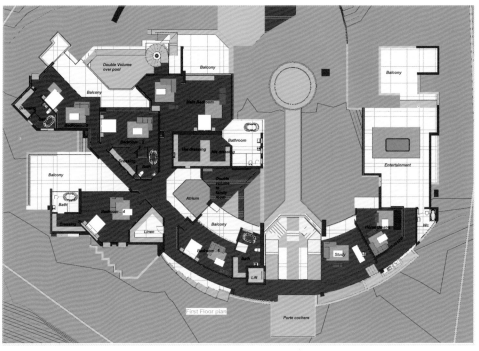

playroom are adjacent to the kitchen, allowing the younger kids to be supervised from the kitchen and family room, and allowing direct access to the bar and barbeque area from the kitchen.

An atrium between the family room and the kitchen allows the family to ventilate and cool the house naturally, without compromising their security, while a roller shutter door drops down automatically when the alarm is activated, cutting the top floor off from the ground floor. The walls to the family room and bar are clad with marble strips, with glass inlays and LED strip lights.

You can jump from the main bedroom into the pool, swim to the gym, swim back and use the steel spiral staircase to go back to the main bedroom, or tip a tipsy friend into the pool from his barstool!

The lanai opens up totally to the outdoor pool with a deck, spilling into a kid's splash pool at the bottom. A basement under the house has parking for about 12 cars, with a view into the pool, and a top — lit art gallery which forms the passage between the garages and the lift.

A feature wall opposite the living areas is clad in stone from Jerusalem, with a tree aloe growing in front of it. The stone comes from buildings hundreds of years old, being demolished in Israel to make space for development. The same stone is used in the dining room, flowing through the glass wall to the outside.

The study is a glass box at the top of the staircase, with a view over the pools at the bottom.

A large playroom is situated next to it, with an intimate home theater and kitchenette, leading to a large balcony with a shaded porch. The main bedroom on the other side of the hall is reached via a gallery looking down into the dining room and out to the garden. The main suite has a small lounge and built—in kitchenette, with a drop—down screen and projector built into the bulkhead. The main bathroom is a study in glass and transparency: the north and east walls are glass and slide open, even if privacy is required, the doors can be left open and the automatic blinds can be lowered, still allowing views and ventilation, but looking translucent from outside.

A large balcony off the main bedroom is partially covered, granting respite from the summer sun, or allowing all fresco early morning coffee or late afternoon drinks, while a staircase to the roof allows views over the surrounding suburb and towards Midrand. The double volume glass enclosure over the pool can be opened from the balcony outside the children's bedrooms, allowing a cooling updraft over the pool. From another balcony the door overlooking the double volume in the family room can be opened, again resulting in a cooling chimney effect to the living areas.

House Moyo

Architect: Nico van der Meulen Architects
Location: Eccleston Rd, Bryanston, South Africa
Site Area: 4,000 m²

The house was designed around three massive trees, one at the tennis court, and the other two on the east side of the house. The view from the hall towards the east is into one of these trees, framed by a large sliding window in the kitchen. You can feed the birds from the kitchen window!

To the north of the lounge, bar and family/ breakfast room is an over—sized lanai and patio, with an eighteen metres infinity edge pool beyond that.

The kitchen, breakfast room and family room lead seamlessly through wall to wall frameless sliding doors onto the patio, with the dining room situated behind the family room, sharing a fireplace with the family room. When the doors are open, the family room, breakfast room, open and covered patio and kitchen become one large area.

Part of the dining room floor is glass, with a view down into the koi pond, and two huge lights from M Square are suspended over the dining room table.

The kitchen is an open plan layout to the family room, breakfast room, and lanai. Behind the black glass back wall a pantry is hidden, and around the corner is a cold room and scullery/laundry.

The basement consists of a gym with spa and a dressing room, a home theatre with a wet bar, a wine cellar/ tasting room with separate areas for white and red wines, a music room and a dance studio/ discotheque.

The wall connecting these rooms has vertical slits lit with fluorescent lights at the back of the slits.

The first floor consists of three kid's suites, with a pyjama lounge with its own mini kitchenette, and an open

plan main suite with a private lounge and a panic room fitted with a kitchenette and desk/ dressing table.

All suites and the lounges lead onto balconies. The main suite has a covered balcony that juts into the family room below, and fireplaces in the private lounge, main bedroom and one of the outdoors on the balcony.

The main bathroom is divided by a fireplace from the main bathroom and built into the same tree as the kitchen below, again with a massive sliding panel opening into the tree.

The toilet and bidet are housed in a dark glass enclosure to cut it off from the open plan suite.

A bridge suspended by stainless steel cable spans across the double volume spaces, with views to north, east and south, and a bridge across the atrium below connects a large balcony to the house.

From the north the house is framed by several beams, with the pond spilling over a three metres wall.

Due to its north orientation, sun control features and cross ventilation, the house can be used almost year round without artificial heating or cooling.

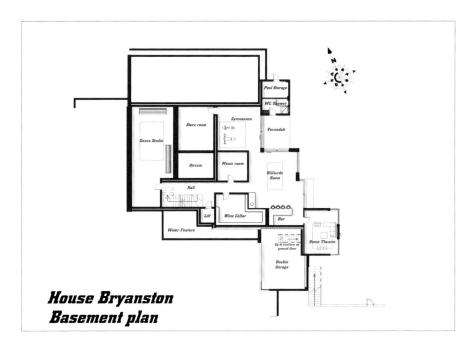

**House Bryanston
Basement plan**

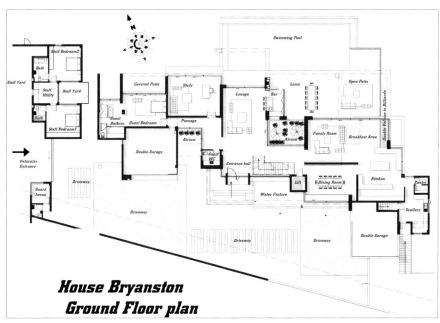

**House Bryanston
Ground Floor plan**

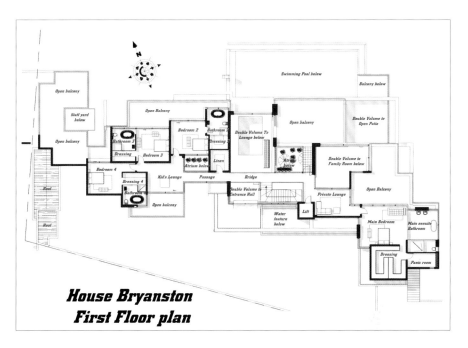

**House Bryanston
First Floor plan**

House Pollock

Architect: Metropole Architects
Location: Durban, South Africa

The site is in the Zimbali golf course estate (Durban, South Africa), and overlooks a vista which includes the rolling green fairway, water features, indigenous forest, and the ocean.

The home, while contemporary in nature, showcases the simplicity of Japanese influence. The clients want a modern 21st century design inclusive of the beauty of natural wood and glass, to allow them to experience as much of the outdoors as possible inside their home. The architects introduced large roof overhangs, big openings and strong horizontal elements into the details of the home, thereby accentuating and highlighting the Japanese undertones. The steep topography of the site, while presenting a challenge, inspired the architects to produce a multi—level, open plan home that projects out of the hillside. The south facing orientation proved a challenge, as ensuring natural lighting into the middle of the house was difficult. This was overcome by the use of an atrium, which serves a twofold purpose: not only allowing the natural light to illuminate the middle of the home, but also separating the resident's area from the guest area.

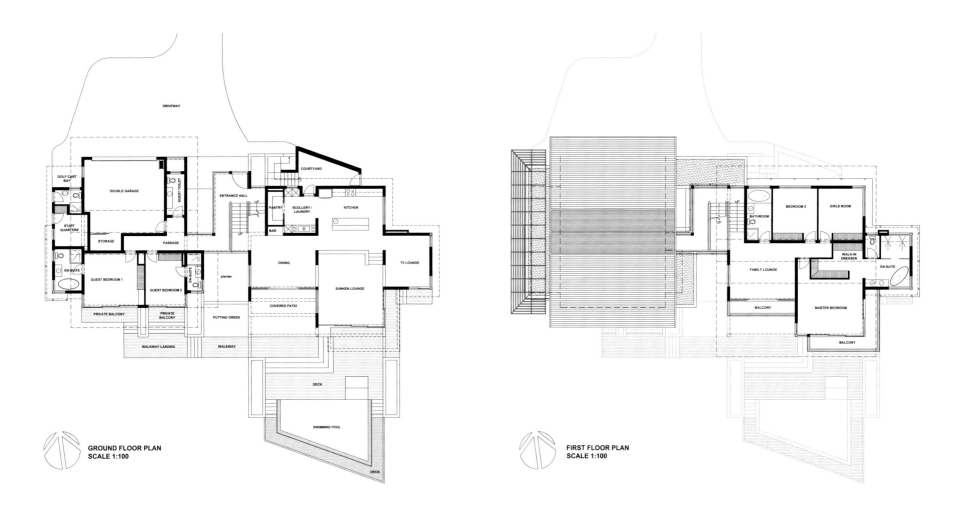

GROUND FLOOR PLAN
SCALE 1:100

FIRST FLOOR PLAN
SCALE 1:100

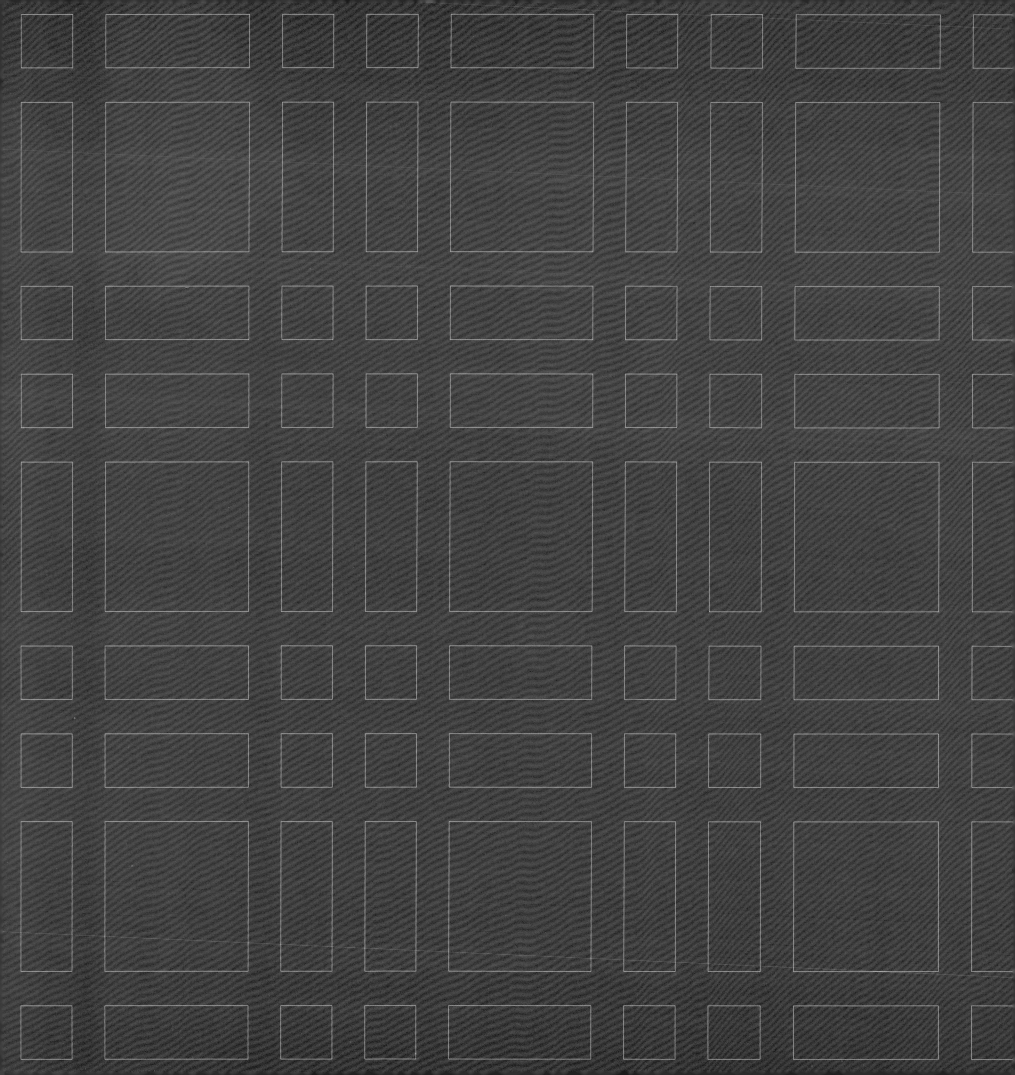

4 Connecting Boxes

Architect: LATO Design (Singapore)
Location: Singapore
Area: 285 m²

The project is a 285 m² two—story semi—detached house for a bachelor in his 30s. LATO Design managed to convince the owner to conserve resources by renovating the house rather than to erecting a new one. Therefore, the issue with the project is how to create a dramatic transformation to the interior spatial quality of the house without major architectural interventions.

"4 Connecting Boxes" of different sizes are being inserted into the house connecting various spaces together.

Box 1 connects the kitchen, dining and living spaces. Intimate dining space is created. The box also becomes a suspended TV console at the living area.

Box 2 is a semi—outdoor terrace space connecting the garden at the back with the living space.

Box 3 connects the master bedroom to the outdoor planter. It creates screening from direct sunlight and also creates a nice cove feeling at the master bed area.

Box 4 at the staircase connects the 1st storey with 2nd storey.

Chengai Timbers are used for timber boxes at outdoors because of weather. Timber veneers are used for timber boxes walls and solid timber strips for timber boxes floor at indoors because they create a more intricate feel. All three are color and tone matched so they look like a whole. Lightings also accentuate the significance of boxes as the main concept of the house.

1ST STOREY PLAN

2ND STOREY PLAN

Sentosa Cove House

Architect: Forum Architects Pte Ltd
Location: Singapore
Land Area: 770m²
Photography: Albert Lim

Sitting on the slope of Mount Serapong, the Sentosa Cove House enjoys unimpeded views of the sea with the Serapong golf course in the foreground. The house is laid out in a linear configuration with the long side running parallel to the sea. A tall slate feature wall perpendicular to the main facade anchors a floating staircase at the east corner.

In order to maximize the panoramic views, the house is ordered into four long horizontal stackable layers with all the major rooms lined up facing east. In what Louis Kahn had famously referred to as the "served and servant spaces", the rooms are clearly organized on plan with spaces such as the corridors, vertical staircases, bathrooms and back—of—house areas located along a spine tucked against the sloping terrain while the feature rooms such as the living quarters and bedrooms enjoy views overlooking the private pool, golf course and the sea beyond.

The approach to the house via the rear sees a solid travertine stone, timber and steel patterned wall. Once inside, the living and dining rooms located on the 2nd storey open up to an unobstructed, elevated view over its verdant surroundings. With light streaming in from the full height glass windows along one side and from the skylight atop the staircase on the other, the marble—tiled living area is an open and delightfully light—filled space in the day. The outdoor terrace encircling the perimeter at the 2nd storey provides an extension of the living quarters to the outside. Descending to the 1st storey, the entertainment lounge and guest quarters open seamlessly to the cool sheltered outdoor patio and infinity lap pool.

The master bedroom and other bedrooms on the 3rd storey have access to the roof terrace on the upper level. A quiet place to come to at the end of the day, the roof terrace is the best place to view the sparkling lights of the Keppel Harbour in the distance.

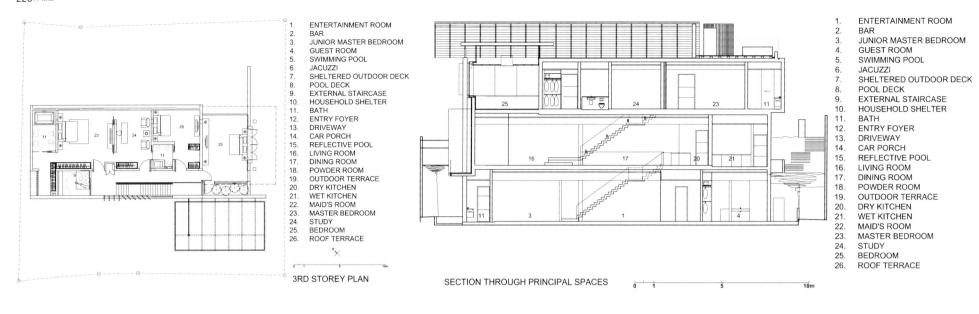

3RD STOREY PLAN

SECTION THROUGH PRINCIPAL SPACES

1. ENTERTAINMENT ROOM
2. BAR
3. JUNIOR MASTER BEDROOM
4. GUEST ROOM
5. SWIMMING POOL
6. JACUZZI
7. SHELTERED OUTDOOR DECK
8. POOL DECK
9. EXTERNAL STAIRCASE
10. HOUSEHOLD SHELTER
11. BATH
12. ENTRY FOYER
13. DRIVEWAY
14. CAR PORCH
15. REFLECTIVE POOL
16. LIVING ROOM
17. DINING ROOM
18. POWDER ROOM
19. OUTDOOR TERRACE
20. DRY KITCHEN
21. WET KITCHEN
22. MAID'S ROOM
23. MASTER BEDROOM
24. STUDY
25. BEDROOM
26. ROOF TERRACE

Sentosa House

Architect: Concrete Architectural Associates
Location: Sentosa Cove, Singapore
Total Area: 520 m²
Photographer: Sash Alexander

The house is situated on a corner lot with two sides facing a water canal. The architectural intent to celebrate this unique waterfront view was achieved by strategically locating a free—standing oval shaped living room that anchored the project on the site, orientating it towards the water. The rest of the house is contained within a fluid and natural form and serves as a backdrop to living room.

The swimming pool is placed within a semi—open courtyard that mediates the living room and the main form. The free—form pool is partially covered by a roof that is reminiscent of a canopy, something closely tied into the idea of tropical living that allows one to exist outside while simultaneously being protected from the elements. The large aperture in the roof, inspired by the Pantheon, allows for natural light, ventilation and precipitation into the swimming pool while simultaneously giving the roof a sense of weightlessness, allowing it to "float" above the rest of the house.

The veranda, a vertical projection of the living room, offers views towards the canal from above, giving one the impression of being at the hull of a boat.

Every architectural decision was focused around the site and the views it provided. With half of the project open to the canal, the waterfront became a focal point around which the house was designed. Each room in the house is orientated to face the water, and both fully exploits and celebrates the views of the waterscape.

The public facade of the project was designed to seem "faceless" for both privacy reasons as well as to provide shade from the strong western sun. The walls seem to "peel" off the building, morphing and transforming as one moves past, before becoming fenestrations that face north.

Concrete's assignment was to design the interior of this weekend house. The aim was to work with the architecture, emphasizing the courtyard and ensuring harmony between the interior and exterior. It was important

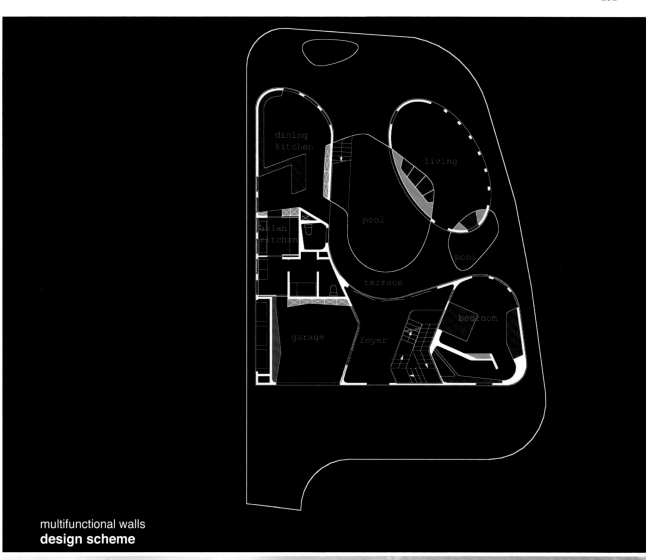

multifunctional walls
design scheme

that the two melded with each other, creating a seamless flow of circulation that did not simply occur laterally from outdoor to indoor, but vertically between floors. In this manner, the entire house becomes an endless playground that one meanders through, always open to views of the water, yet intensely private at the same time.

This seamlessness has been further realized in the project's furniture and finishings. The use of Corian allows for a continuous finish that flows through the entire house, allowing the finishing to transform into one large piece of "furniture". The Corian allows beds to develop into shelves, which in turn become cupboards, vanity displays, and kitchen countertops. Without the need for edges, the material wraps the building on the interior and draws the individual from one room to the next, perpetuating the idea of an unbroken flow of movement.

The garage, a space one would usually consider separate from a house, or belonging to the outside more than the inside, is instead made a cohesive part of the interior. It is unenclosed, separated from the foyer only by a glass wall. While most garages are hidden, this one is designed as a private display case, a mini automotive museum of sports, where a wall of 1:24 scale cars serve as an impressive backdrop to a 1:1 Ferrari or Lamborghini.

The oval living room is the penultimate realization of the seamless flow of movement. The swimming pool seems to slip into the living room as its mosaic tiles cover the zone where the geometries of both spaces begin to interact with each other. The lighting scheme of the room allows it to be lit in various colors throughout the day, adjusting the atmosphere of the room to suit the mood and preference of the user. In this way, the space can either blend into the environment or become a single glowing object in the dark.

Villa Overlooking the Sea

Architect: EDWARD SUZUKI
Location: Hayama, Japan
Site Area: 2,620.84 m²
Photography: Yasuhiro Nukamura

This is a vacation home expected to become a permanent house in the future. Located about an hour's car ride away from downtown Tokyo, it is located in the sea resort of Hayama in the Shonan region, adjacent to the more well known, old city of Kamakura.

Sited overlooking the sea on the south, the design intent is to capture as much of the scenic panorama and to bring it into the house. Behind the house on the north is a hill clad with wild bamboos. Sandwiched in between the richness of the sea and that of the hill, the design is to open the house as much as possible to the natural outdoors and to become a part of it.

The plan is basically a multiple of six—by—six meters concrete grid in three stories. The basement approach level is predominantly entrance, garage, and storage. The first floor is sleeping quarters, and the second, uppermost level, is living, dining, and kitchen. The second floor is basically one large room zoned functionally and separately by the structural grids.

Interior floor finish for the second floor is laminated bamboo, much used by the architect's office recently, as the material is very fast—growing and therefore more eco—friendly than wood. The bedrooms floor is carpeting to allow softer material for the legs. The exterior is stucco clad over exterior insulation to prevent interior condensation and subsequent mold.

Extending the full length of the house on the south sea side is a three—meter wide terrace balcony acting as an interface between the outside and the inside. On sunny, non—cloudy days, one may be able to view Mt. Fuji in the southwest in all its glory.

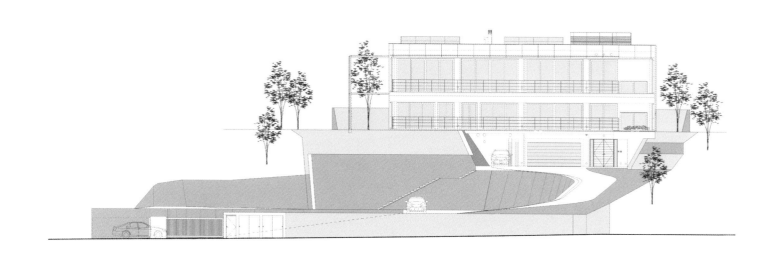

0 1 2 5M

ELEVATION

VILLA OVERLOOKING THE SEA

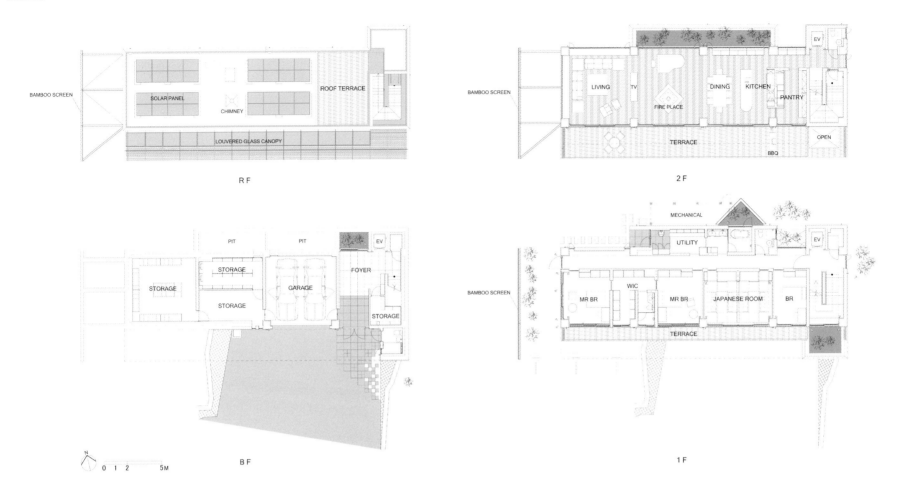

RF

2F

BF

1F

House of Maple Leaves

Architect: Edward Suzuki
Location: Karuizawa, Japan
Site Area: 6182.97 m²
Photography: Yasuhiro Nukamura

This is a villa situated about an hour's rapid train ride from Tokyo in a mountain resort of Karuizawa, Japan. It is basically a wooden structure with the peripheral balconies in steel. The design is adapted from passive energy principles applied in the world—famous Katsura Imperial Villa in Kyoto. The only artificially operated system is the radiant hot—water heating embedded in the floor, running the entire length of the peripheral fenestration with heat pump air conditioning units as supplements just in case.

Legal and binding design guidelines in the area require a roof slope of 1:5 minimum, eave length of 500 mm minimum, and a limited selection of exterior coloring.

SAZAE's House

Architect: Akasaka Shinichiro Atelier
Location: Sapporo,Hokkaido,Japan
Site Area:181.81m²
Photography: Koji Sakai

This client having three family members, a couple in their thirties and a pre—school young daughter, decide to live with the husband's mother and grandmother, brother and sister in their twenties, as they build a new house. They have seven family members of four generations, which is rare to see these days. When the architects consider overlaying such family structure with the lifestyle of a young client, the architects thought the relationship and distance between seven members are more like the dormitory or lodging house where people gather having different lifestyle and schedule, rather than the ordinary resident.

There are two passages intercrossing in different levels in the center of the house, facing five small private rooms and also a water supply control system.

By taking the volume for the passage (common area) as large as possible in the limited area, the architects aim to design a house where seven people can be connected comfortably, having each member to be attracted to the common area rather staying their rooms.

The passage that refracts with the reflection of light, the spiral staircase located at center of house, and the large insulated window located in the end of open ceiling, all function as a clue that each family member discovers their own comfortable space.

Now, dining table and TV are placed in the first level of the passage, functioning as a place where the entire family gathers. The second level is for young members to read magazines and listen to music. Through the indoor windows installed in the private rooms, the presences of each person inside are permeated to the passage.

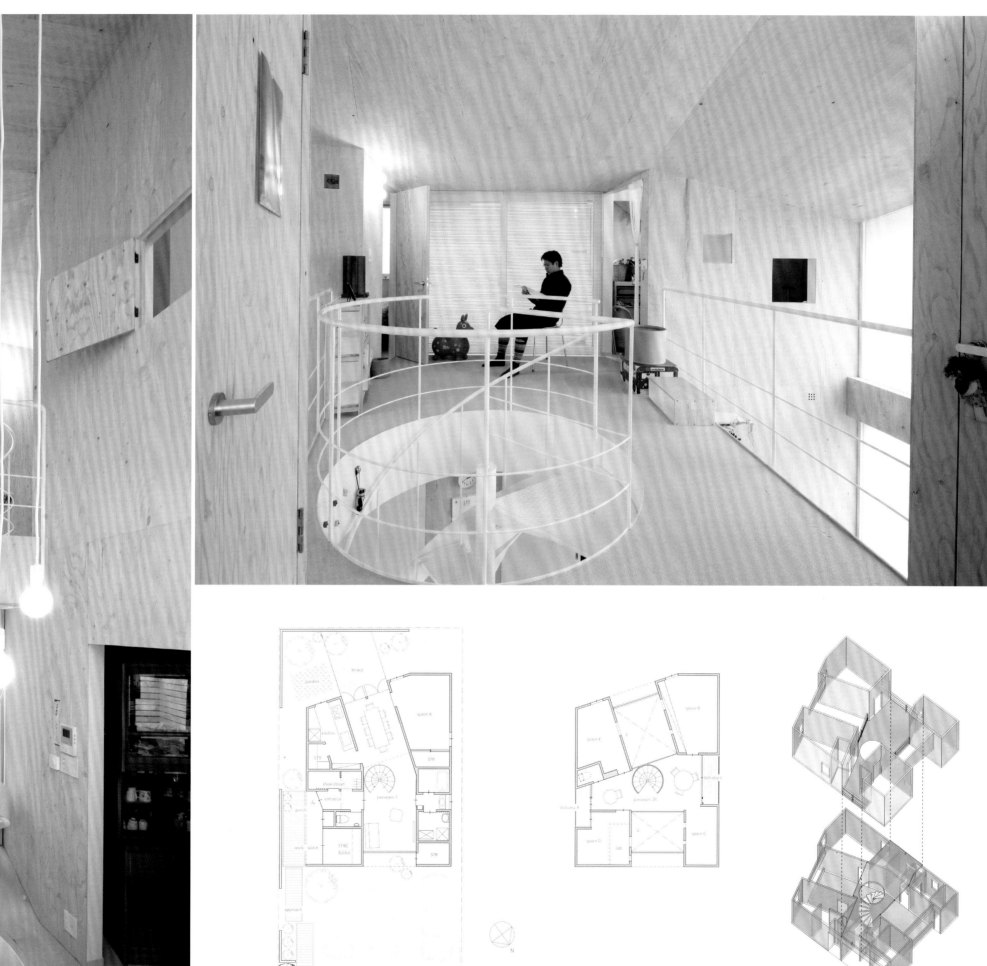

Hanil Visitors' Center and Guest House

Architect: BCHO Architects (Byoung Soo Cho)
Location: Chungbuk, South Korea
Land Area: 3,957 m²

The Purpose of this project is to educate visitors about the potential for recycling concrete. In Korea concrete is one of the primary building materials so it is imperative that the architects begin to re—use, the otherwise wasted, concrete as buildings come down and are replaced. The Hanil Visitors' Center is an example of how to reuse this material in different types of construction, casting formwork types as well as re—casting techniques. Concrete has been broken and recast in various materials creating both translucent and opaque tiles. The displays will continue to evolve and change at the Hanil visitors' Center as new techniques are developed. The gabion wall, with fabric formed concrete which constitutes the main facades of the building, was erected first, and the concrete left over from it was recycled in the gabion cages, on the rooftop for insulation from sun, and as a landscape material at the street and around the factory.

The site is located to the westernmost part of the factory, adjacent to Mt. Sobaek National Park. The existing land had been changed much to facilitate the movement of trucks to the cement factory. First of all, the architects tried to restore the damaged original mountains and forest. In order to revive the landscape, the architects brought in earth to fill the courtyard between the two buildings. The flow of the mountains from the west leads to the reception and cafeteria in the inner courtyard of the building. In the in—between spaces the architects allowed people to experience the mass of the building while watching the building shift around its central courtyard.

While following the linear placement and movement of land and earth, the architects came up with ideas for the new building facade. The architects applied canvas—like concrete walls to the east facade, evoking images of the adjacent forest. There are four openings in the eastern wall and long vertical

'C-C' Section 0 5m

East Elevation 0 5m

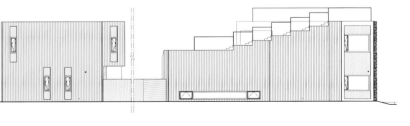

West Elevation 0

windows have been created in their in—between spaces. Through the windows, one can see how concrete is produced at the factory. Behind the two larger openings, one can see the courtyard of the Hanil Visitors' Center and the cafeteria next to the courtyard, which is encircled by a water garden.

By recycling waste concrete generated in erecting the eastern wall, a new concrete wall was erected on the opposite side. The footing used in erecting the fabric formed concrete wall was cut into pieces of 10 cm—20 cm and put into the gabion wire netting to be recycled as exterior finishing material for the southern facade. Recycling of waste concrete not only has eco—friendly and cost benefits, but also offers an antique feel, as dust and moss gather on the concrete with the passage of time.

In order to keep the physical properties of concrete intact and simultaneously express gentle curves, the fabric—formed concrete wall was developed in collaboration with C.A.S.T., based in Manitoba, Canada, after much research and consultations. Concrete moulds were created on the footing; curved forms were set using pipes, and high—strength fabric was placed on top like a mould. After embedding connecting fittings, concrete is poured, and pre—cast concrete is lifted so that it could be installed at the external wall in the east. While producing a non—bearing concrete wall with convex and concave curvatures, the architects conducted a variety of experiments, departing from stereotyped notions.

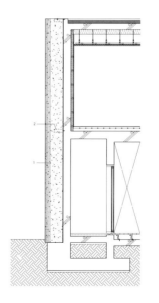
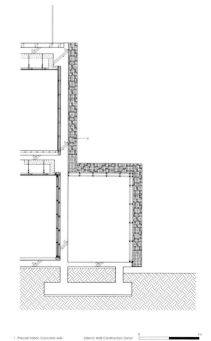

Floating House

Architect: Hyunjoon Yoo Architects
Location: Geonggi—do, South Korea
Area: Main House 195.52 m², Guest House 59.5 m²

The given land has beautiful scenery facing the South Han—River in the north. In the meanwhile, lots of restaurants and motels are seen in the south and decadent neon signs are seen from here to here creating visual pollution.

The clients are a couple in their fifties and sixties who run an elegant Korean restaurant which was built in a traditional Korean style. They want to build a house for them in the given land about five minutes away from the restaurant. They want to overlook the river from a high level, at the height of the third floor. His close Feng Shui specialist had advised them not to make a door facing east and not to let people live in the southwest. They want to have a study room, a barbecue place, and a guest house for their daughter's family who sometimes visit them. They also want to make a big yard to play and jump about and a low—leveled swimming pool for their grandchildren. The wife wants a design that gives a heavy and magnificent feeling.

The house is located in the north as far as it can be for a big sunny yard in the south. The guest house is located in the south so that the restaurants and motels are not seen from the yard. The guest house is built as if it are the fence to be located as far as it can be in the south. It looks like it is put inside the thick fence. To view the river from the yard and the guest house, the main building is leveled up to be a piloti. The roof of the main building is made as a plain roof and there is a roof garden in which people can overlook the river from high level. As a result, the river is viewed from every part of the house.

To make the house look as big as possible, a one meter wide balcony made of the same materials as the main building is built at its four sides. The regulation defines that a balcony with one meter width is not included in total floor area. By doing so, the interior looks as if it keeps expanding when it is seen from inside the building. Also, since the whole building is lifted from the floor for one floor's height, the pilotis space could be more expanded. Not to make the pilotis look big—headed like most piloti space with small and minimal columns, the columns

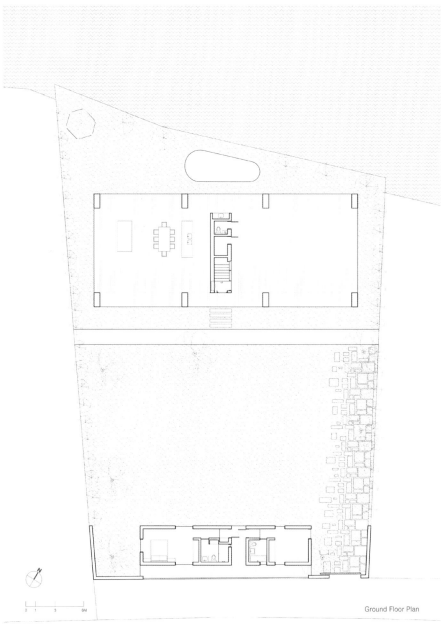

Ground Floor Plan

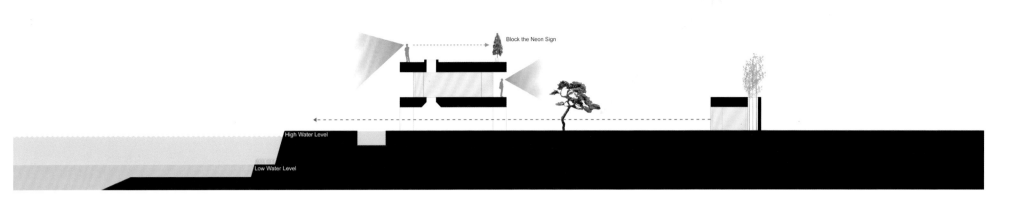

were made thicker than necessary and a waterspout was installed inside to make the first floor look like a heavy rock. Black concrete was used for a magnificent style, creating the feeling of tiles used in traditional Korean—style houses. As a result, the house created an image which is similar to the Korean—style restaurant of the clients.

The plan can be described as the "space within space". The service space, such as a bathroom, a dress room, and a kitchen, is concentrated in the middle and compactly distributed, and a bedroom is placed in the east, a living room in the west, and a corridor in the outskirts. To make people in the house feel the space bigger than real, a circular traffic line is adopted instead of one main traffic line. The circular traffic line is made with a minimum width to differentiate the intensity of space, but the space does not look small because the traffic line extends to visually meet the terrace. Since the stairs that lead to the roof are made as if they are inserted to the second floor, the staircase does not protrude on the roof. The staircase itself plays a role as a well of light, transmitting light to the middle level of the house.

Architecture is to design relationship. There are three kinds of relationships: physical relationship, visual relationship, and psychological relationship. "Physical relationship" is the relationship in which a person looks at an object and actually he/she can reach that. The example is space where a person can communicate with a gate, a corridor, a road, or a bridge. "Visual relationship" means the relationship in which a person can look at an object but he/she cannot reach that. When a person looks at the ground across the river that has no bridge, and he/she communicates with that place through a window from a place without door, it is the visual relationship. "Psychological relationship" is the relationship in which a person cannot go over or look at an object, but he/she knows that there is the object. When a person is in a place without a window and he/she knows that there is a room beyond the wall, it is the psychological relationship.

In this house, these kinds of relationships are created through traditional "rooms," such as living room, bedroom, roof, staircase, kitchen, or bathroom. For example, the study room and the kitchen are divided by a staircase but connected through a small window. The bathroom is connected to the sky through the window at the top and connected to the staircase that leads to the roof at the third floor through the window. Also, a person can look at the river through the study room while he/she is taking a shower or bath. In the living room, there are same—sized windows at the top and bottom to provide a wormhole—like feeling to the parallel structure. Through this mechanism, a visual relationship is formed at each floor. The windows and doors here and there are devised to create more varied relationships in the simple mass—shaped house.

Island House

Architect: IROJE KHM Architects
Location : Gapyunggun, Gyeounggi—do, South Korea
Site Area : 872.63 m²
Photography : JongOh Kim

This site, floating on river and confronting the graceful landscape, was strongly recognized as a part of nature, from the first time that the architects met. From then, the architects started to visualize "the architectural nature" as a place of recreation.

While maximizing the efficiency of land use, the leaner concrete mass, that cherishes the courtyard which is filled with the water and the greenery, was laid out on this site along the irregularly formed site line. This courtyard is "the architectural nature" and a central recreation space as extended river that communicate the river and architecture.

The whole part of the step typed roofs, are directly linked to the bed rooms in upper floor is moving upward with various level. Finally, the stepped roof gardens are linked to both sides of the inner court where swimming pool is.

The inner court which is filled with the water, flower and fruit, and the whole of the roof gardens are circulated as the continuous landscape place that is the place as "architectural nature" in concept.

Naturally, all of the rooms inside this site—shaped—mass are laid toward the picturesque landscape to enjoy the graceful scenery surrounded this site. The huge panoramic view, framed with sloped ceiling line that is composed with the lines of stepped roof gardens and the bottom line of the inner court, is the major impression of inner space of this house.

Skip floor plan of inside of this house produced various dramatic spaces.

Asia

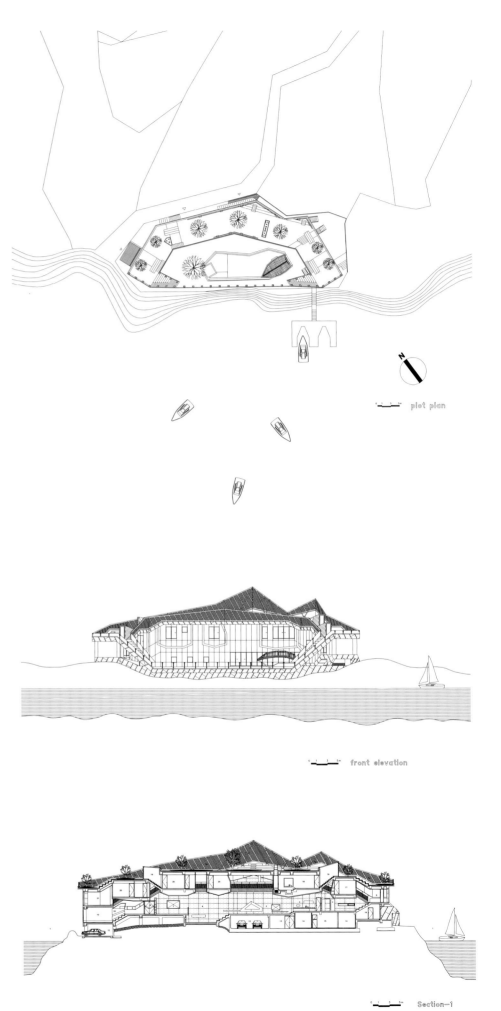

plot plan

front elevation

Section-1

As "the architectural nature", floating white polyhedral masses that have the built—in bamboo gardens, produced the various stories of vertical space

The shape of the mountain type composed of irregular polygonal shaped concrete mass and metal mesh is designed to harmonize with the context as "the architectural mountain".

There is the intention to be a part of the surrounding context that consists of the river and the mountain.

As a result, this house is to be "the island house" as an "architectural island".

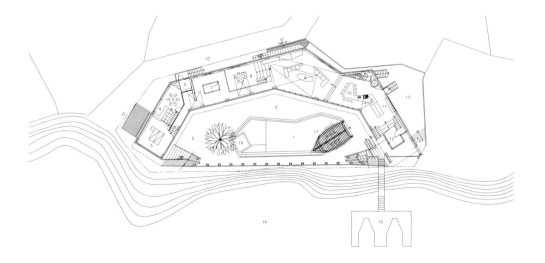

first floor

Purple Hill House

Architect: IROJE KHM Architects
Location: Gyeounggi—do, South Korea
Site Area: 554 m²
Photography: JongOh Kim

This scenic site is a part of the residential area developed in the natural greenery area of the mountain Gwanggyo and is located at the entrance area to trail to mountain.

From agony about the direction to which major living space sees, the plan of this house started.

After all, the direction to the opposite mountain that will be preserved in the natural green space forever was selected.

But this charming selection that introduces the permanent natural landscape brought the problem, bad solar condition, that the major living space is faced to northwest sunlight.

And so, the architects must make the solution to absorb the southern and eastern sunlight into the inside of the house.

"Glass box of light" that will absorb the useful sunlight was introduced to inside space of the various levels and these boxes which are planted with flower and fruit function as "floating glass garden".

Actually, the architects solved the problem and own the two conditions that are the introduction of landscape and ecological environment and naturally have been reached to the complex solution that all the rooms have their own garden.

The dynamic, unrealistic interior space which is produced with several floating glass boxes over the major living space that is opened to the vertical height of 3 floors, will be the symbolic impression of this house.

The architects expect that this landscape architecture will coexist with the surrounding context for a long time, as if a small purple hill...

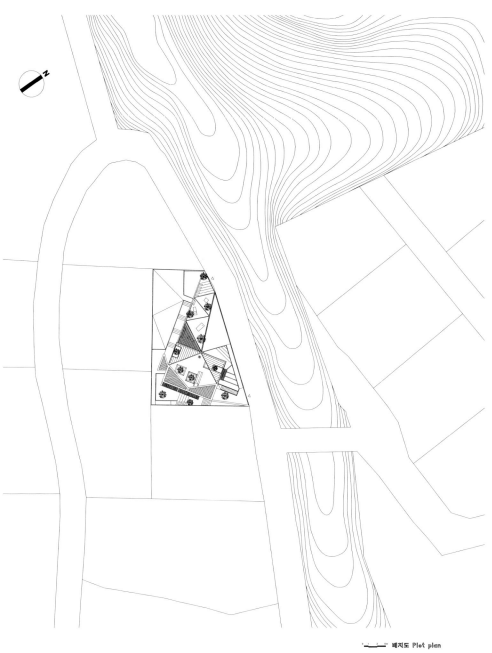

배치도 Plot plan

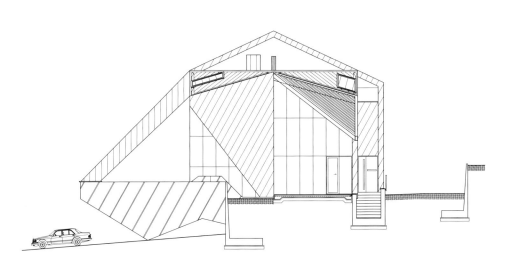
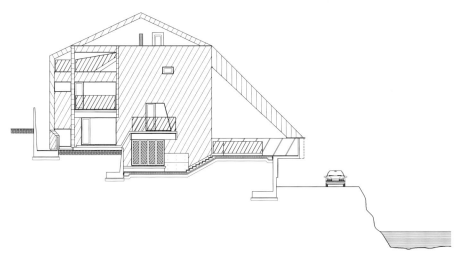

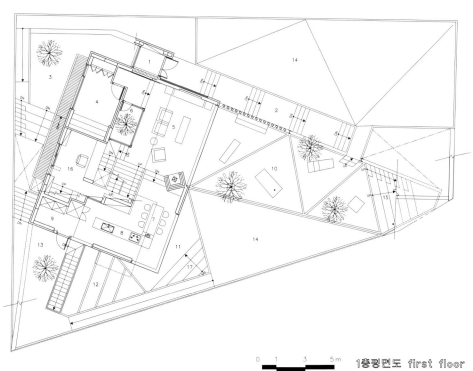
1층평면도 first floor

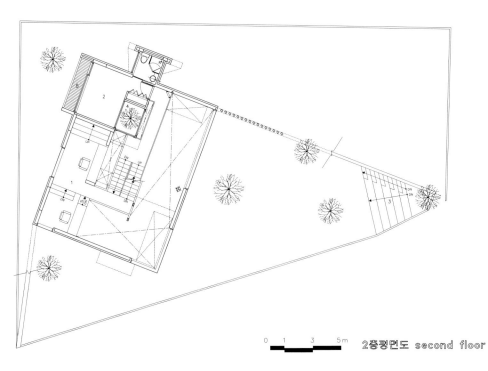
2층평면도 second floor

Contemporary Bauhaus on the Carmel

Architect: Pitsou Kedem Architects
Location: Haifa, Israel
Site Area: 1,000 m²

A private residence built in the center of a historic avenue and at the very heart of Haifa's French Carmel neighborhood.

The avenue is studded with a number of residences designed in the Bauhaus style. The Bauhaus style gained its hold in Israel in the wake of international styling trends and is a ornament free design style, both simple and down to earth.

In this design, the architects has expressed his own, localized interpretation for free planning in which there is a spacial continuity achieved through light, appearance and movement and the placement of secondary spaces around one, large and open central space. The architects has succeeded in creating the experience of continuous, intimate and defined spaces with different levels of symbiotic, mutual interaction with the central space and yet without detracting from the overall understanding of the structure. Despite the intensification of the residences central space which finds expression in a double sized open space reaching the entire height of the building with one completely transparent facade facing the direction of the courtyard, through the use of controlled and restrained formality and the use of materials with no external facings, the designer has succeeded in showing his belief that it is possible to create a residential space of quality and timelessness.

 The home is, as said, designed around a wide, high public space that constitutes the connecting point and provides a view of all of the home's different wings as well as to the central courtyard and the pool. In order to further strengthen the impact of the central space it has been coated with exposed concrete panels and a large library on the wall as a central motif. A large, ribbon window allows light to enter deep into the space,

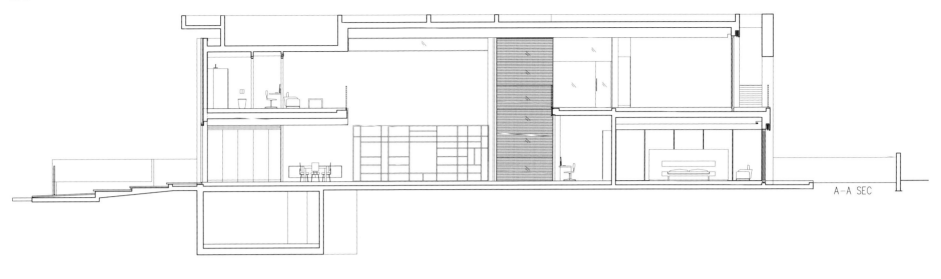

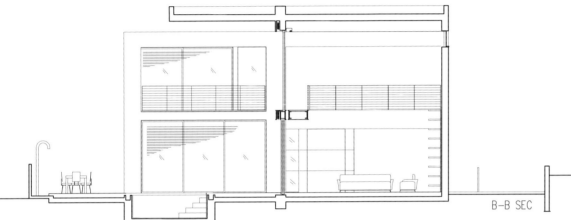

creating movement and dynamism on the central wall. The architects has covered all of the structures spaces with an expansive roof which appears to be suspended, weightless in the air and floating effortlessly with no apparent means of support. The roof frames and consolidates the various parts of the structure with the apparent dissociation between the roof and the building creating an impressive, formal dialogue.

House R — Hasharon House 1

Architect: Sharon
Location: Kfar Yedidia, Israel
Site Area: 550 m²
Photography: Amit Gosher

The house was planned for a young family with 3 children, on a lot with a wide open view to the back, overlooking agricultural fields.

After giving a great amount of consideration to the functionality of the space and the movement scheme, the architects have decided to build a split level house, placing the master bedroom half way between the public area and the children's space.

The scheme evolved into an L—shape plan with the MBR placed in the middle, outside the L, being surrounded by its 2 wings.

Since basement was not part of the requirements, we decided to hang the MBR as a cantilever, thus freeing the space beneath it, enlarging the garden and allowing interesting views across the lot.

The building code allows a space to be under 180 cm height, if it's not to be included in the meters you pay taxes for, so the height is 179 cm.

Using the same methods, an entire wall was hung between the living room area and the staircase \ kitchen \ dining room.

Wanting to focus the view towards the outdoors as much as possible, and also to create an informal, inviting living room, the architects needed to place the TV screen in a manner that would not be turned away from

the scenery, and the floating wall was the answer, maintaining a flow throughout the floor.

One of the main goals is to design a house that is easy to live in, a lot of emphasis is given to storage, most of it hidden in the walls. The architects have used a 130 cm tall / 70 cm deep space beneath the MBR cube for storage, washer and dryer, and mechanical systems.

A study is placed on the ground floor in an enclosed room, while a second, open study is placed on the children's 1st floor.

The local weather allows the outdoor kitchen to be very useful during most of the year. The pergola is able to be open to the sky allowing shade, or closed to rain, using electrical shutters that are either vertical or horizontal.

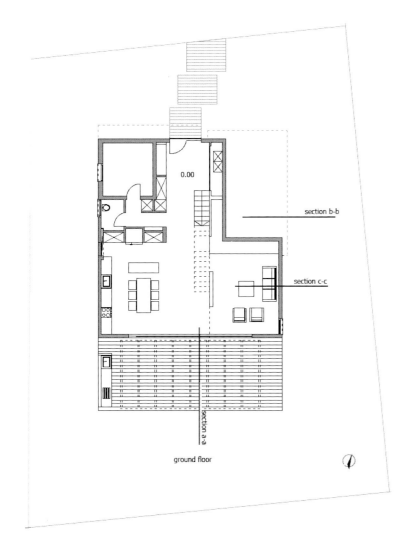

ground floor

House E — Haruzim House

Architect: Sharon
Location: Haruzim, Israel
Area: 300 m²
Photography: Amit Gosher

The house is built on a very long, narrow and sloped lot of 940 m². The proportions are one of the generators for the design. The clients want a large, spacious house on one hand, while maintaining the intimacy and the sense of a warm family life on the other.

The client's dream is clearly defined: strong connection between indoor and outdoor, huge patio as a main feature, connected to most of the rooms. A house that will not be too flashy for its surroundings. And if possible, a plan that would deal with their inherent mess. "We need you to help us to be more organized."

The solution is to plan the house one level, which creates intimacy despite the size. The house is organized around a central courtyard embraced by the public space and the children's rooms.

The house is based on the longitudinal axis as private area and an open square space 120 m² as public area. A dramatic long corridor is lighted by a "light—fall" — a sky light window devised to also let hot air out.

The architects dealt with the subject of an organization and order using a huge piece of furniture designed as multi — functional storage, located in the central open space, accessed from four sided. One of its sides hides the steps leading down to the basement and has a niche displaying a collection of sculptures.

Concrete is one of the most common materials here, and is used in a unique way—thin fiber cement surfaces form the kitchen countertops and wrap the island, coat the window sills, the surfaces in bathrooms and garden benches.

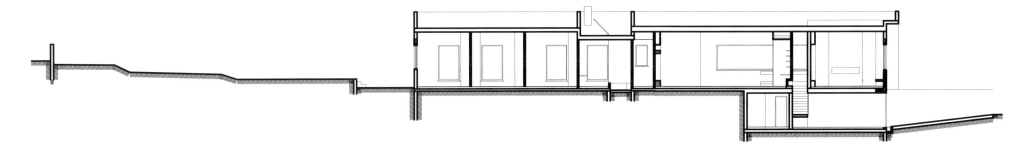

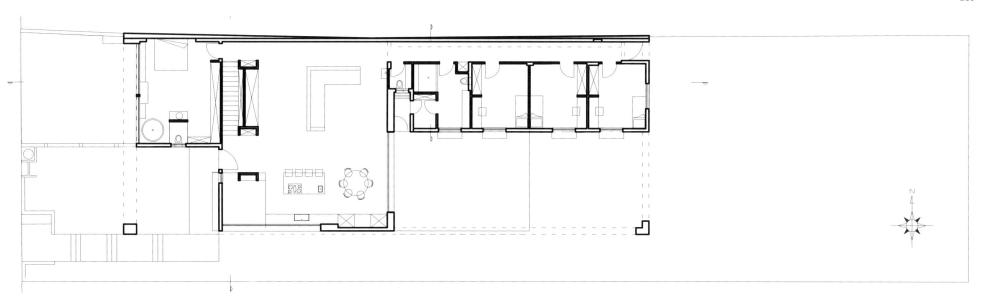

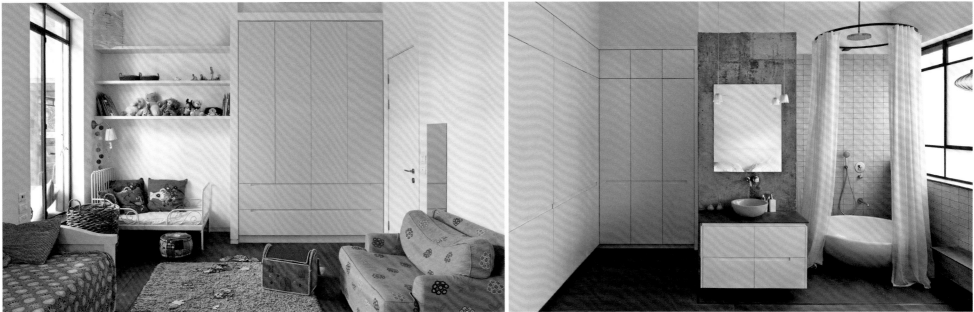

Kibuts House

Architect: Sharon
Location: Kibuts Lehavot Haviva, Israel
Site Area: 430 m²
Photography: Amit Gosher

The cube, brick house is planned for a family of 2— father & son. A large open space on the ground floor contains the entrance, kitchen, dining and living rooms. Double height space above the dining table allow the stairs to go around it. A separate area contains the study and bathroom. The master bedroom upstairs has an open plan with free standing colorful walls which define the bed, bath and closet space. A detached pergola keeps the cube untouched. The garden is designed for water saving.

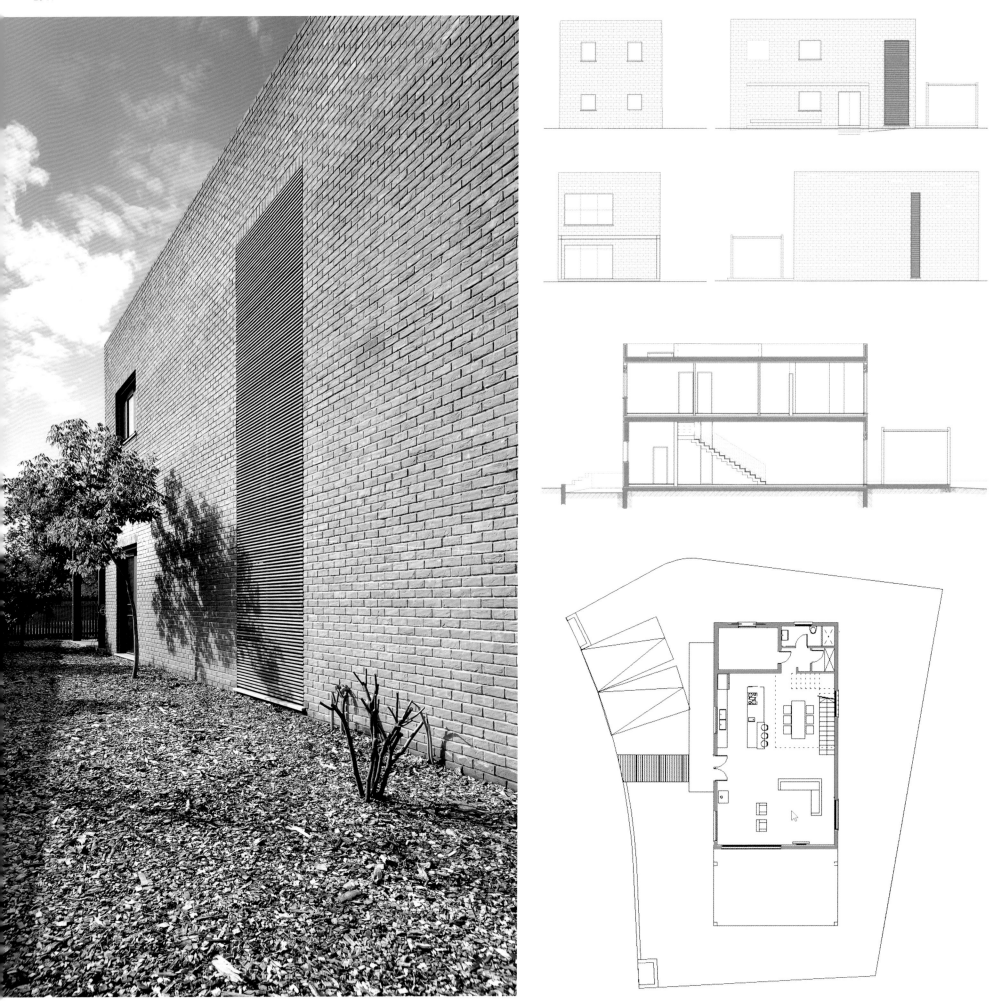

Ribbon House

Architect: AGi architects
Location: Kuwait
Photography: Nelson Garrido

The organisation of this dwelling is based largely on the client's need to accommodate 3 generations of the family into the same house. Hence, separate wings for the different family members are located on various levels of the house to achieve privacy from each other. These are all organized around the central courtyard that houses the pool. The vertical circulation also revolves around this courtyard space, as one goes up the stairs the landing extends into the landscape of the courtyard below.

With the growing programmatic needs of the client, the facade becomes an important element in the design. It is used to unify the complexity of the extensive program. Horizontal bands wrap around the house, while the 2nd floor is set back in order to break down the scale of the project. These bands are angled differently on the various floors to achieve a visually dynamic expression.

GROUNDFLOOR

1. Hall / Main Entrance
2. Guest Living
3. Guest Dining
4. Dewania
5. Family Living
6. Kitchen
7. Toilet
8. Lounge
9. Bathroom
10. Bedroom
11. Storage
12. BBQ
13. Pool

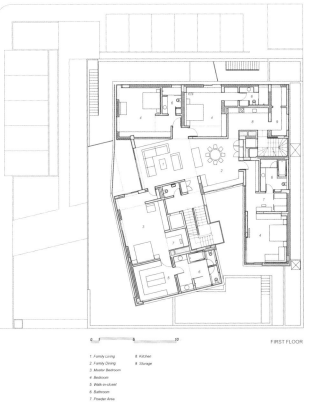

FIRST FLOOR

1. Family Living
2. Family Dining
3. Master Bedroom
4. Bedroom
5. Walk-in-closet
6. Bathroom
7. Powder Area
8. Kitchen
9. Storage

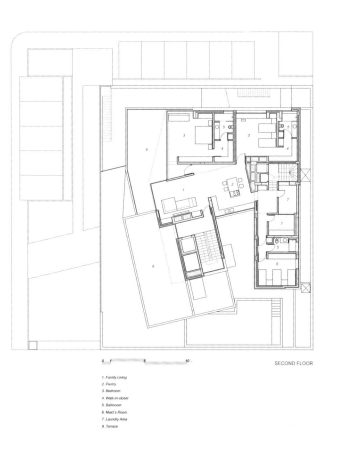

SECOND FLOOR

1. Family Living
2. Pantry
3. Bedroom
4. Walk-in-closet
5. Bathroom
6. Maid's Room
7. Laundry Area
8. Terrace

North America

Nakahouse

Architect: XTEN Architecture
Location: Los Angels, California, The United States
Floor Area: 251 m²
Photography: Steve King

Nakahouse is an abstract remodel of a 1960's hillside home located on a West facing ridge in the Hollywood Hills, just below the Hollywood sign. To the South and West are views of the Beechwood Canyon, to the East is a protected natural ravine, with a view of Griffith Park Observatory in the distance.

The existing home was built as a series of interconnected terraced spaces on the downslope property. Due to geotechnical, zoning and budget constraints the foundations and building footprint are maintained in the current design. The interior is completely reconfigured however, and the exterior is opened up to the hillside views and the natural beauty of the surroundings. A large terrace is added to link the kitchen/ dining area with the living room, with a steel stair leading to a rooftop sundeck. Terraces are also added to the bedroom wing and the upper master bedroom suite to extend the interior spaces through floor to ceiling glass sliding panels that disappear into adjacent walls when open.

The exterior walls are finished in a smooth black Meoded ventetian plaster system, designed to render the building as a singular sculptural object set within the lush natural setting. A series of abstract indoor—outdoor spaces with framed views to nature are rendered in white surfaces of various materials and finishes; lacquered cabinetry, epoxy resin floors and decks and painted metal.

The contrast between the interior and exterior of the house is intentional and total. While the exterior is perceived as a specific finite and irregular object in the landscape the opposite occurs inside the building. Once inside the multitude of white surfaces blend the rooms together, extending one's sense of space and creating a heightened, abstract atmosphere from which the varied forms of the hillside landscape are experienced.

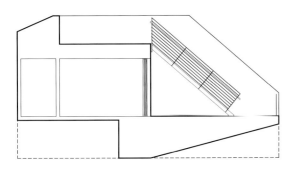
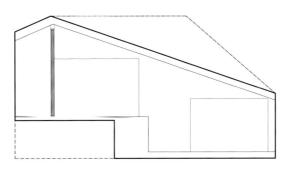
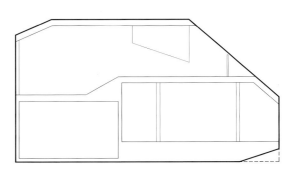

SECTION A-A

SECTION B-B
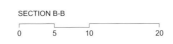

SECTION C-C
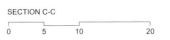

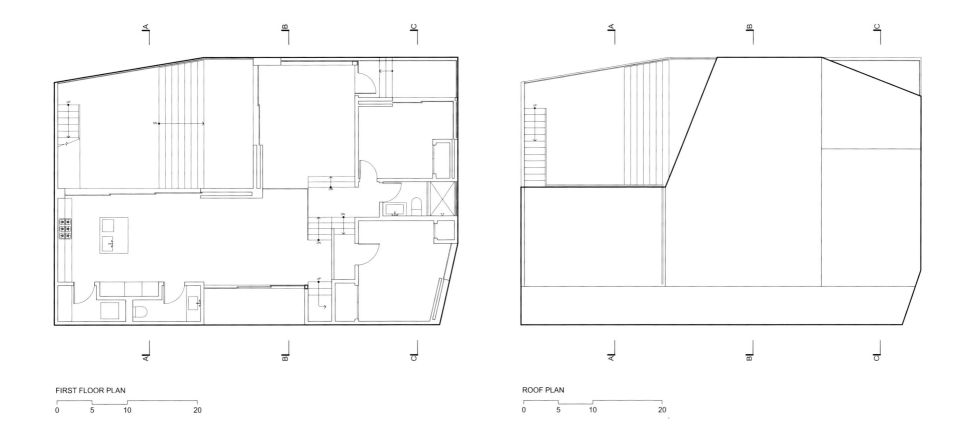

FIRST FLOOR PLAN
0 5 10 20

ROOF PLAN
0 5 10 20

Villa Allegra

Architect: Oppenheim Architecture + Design
Location: Miami Beach, Florida, The United States
Area: 836 m²
Photography: Eric Laignel, Laziz Hamani

Multiple rooms, both interior and exterior, have been added to a non—descript, one—story home transforming it into a receiver of Miami's tropical climate. While the effect is striking, minimal alterations are made to the existing structure. The house is entered through a 20'x30'x30' volume where a reflecting pool and oculus align to activate the space with reflection and luminance. A large room organizes the house into private and public realms. Tremendous spaces with oversized windows overlook the pool and canal. A 60'x20'x20' volume, at the rear of the home provides enclosure for outdoor living. A large circular column contains an outdoor shower open to the sky. The second floor contains a secluded courtyard garden, off the master bedroom, for private activities. The project provides a flexible infrastructure for the participation and enjoyment of the pleasures of life.

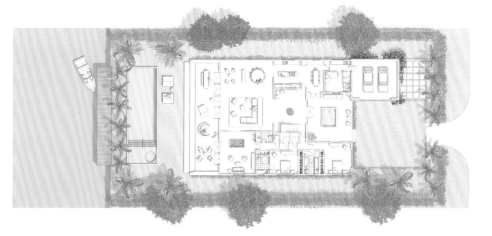
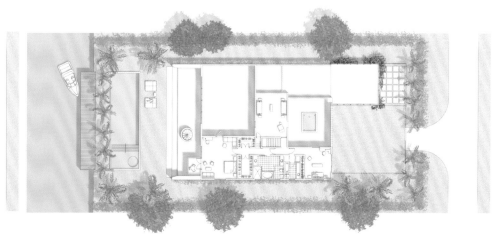

Casey Key Guest House

Architect: Jerry Sparkman, AIA, NCARB, TOTeMS Architecture, Inc.
Location: Casey Key, Florida, The United States
Land Area: 17,500 m²
Photography: William S. Speer, Chroma, Inc, George Cott Photography,
Greg Wilson Group

The project consists of a small single family guest house set within a mature oak hammock, located on a barrier island on Sarasota bay. The narrow island is approximately 600' wide at the project site, and spans between the Gulf of Mexico and Sarasota Bay. Distinct native ecologies of this subtropical environment are clearly evident. From east to west they include: shallow bay waters, mangroves, oak hammock, sand ridge, dune, beach, and Gulf.

Positioned at the east side of the oak hammock, as a transition to the mangrove ecology, the wooden structure is inspired by two elements. First, the Owner's one sentence program which reads, "...respect the land, and the rest will follow".

The house is located in a highly regulated FEMA velocity zone which requires elevated floor levels supported on pile foundations. To preserve the health of the oak hammock a specialized mini—steel pile foundation system was designed to avoid root disturbance and minimize sub grade impact to the live oaks. As a result, all existing trees were preserved.

The second design goal reflects the character and influence of the live oaks. Limbs shaped by the prevailing coastal winds from the west, provided inspiration for the shaping of the structure. To achieve this goal, glulam beams were utilized to enfold the structure around the space. Reflecting the arching quality of the live oak limbs, curved, laminated pine beams anchored at their base to the elevated concrete slab, curve over the entire space, blurring the distinction between wall and ceiling.

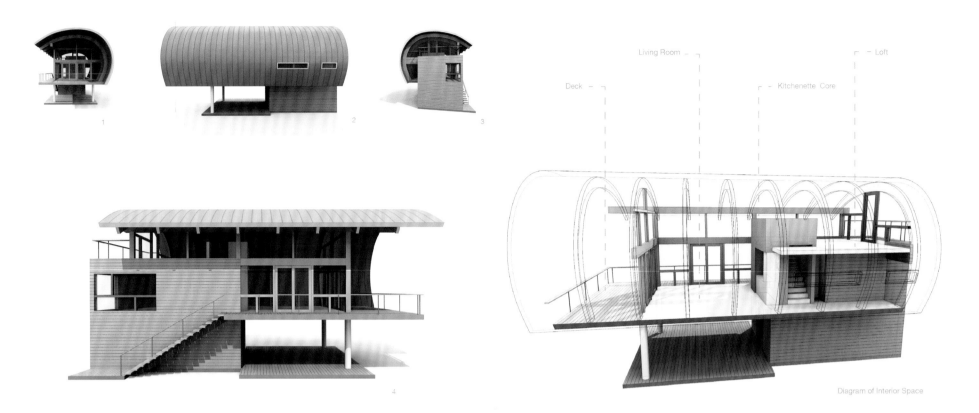

Diagram of Interior Space

The Owner's request for a "house in the trees" consists of a small program, including: one bedroom, one bath, a living area, a kitchenette, and a loft/sleeping area. The spaces are organized to provide privacy between the neighboring property to the north, while offering broad views of the oak hammock to the south and west, and Intercoastal Waterway to the east. The ground floor includes a kayak storage space, and a covered Ipe deck. The first elevated floor consists of a double height living space, kitchen/HVAC/stair core element, and small bedroom. The loft interior, defined primarily by the glulam beams, and tongue and groove cypress siding, inadvertently alludes to the aquatic bay environment, and wooden boat hull construction.

The design is intended to evoke an organic architecture, one that is influenced by, and reflective of its site.

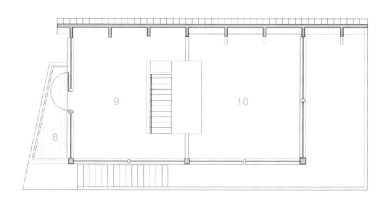
Loft Level Plan

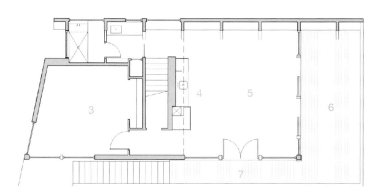
Main Level Plan

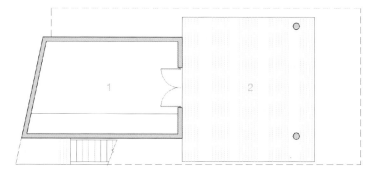
Ground Level Plan

Open House

Architect: XTEN Architecture
Location: Hollywood Hills, California, The United States
Total Area: 697 m²
Photography: Art Gray Photography

The Openhouse is embedded into a narrow and sharply sloping property in the Hollywood Hills, a challenging site that led to the creation of a house that is both integrated into the landscape and open to the city below. Retaining walls are configured to extend the first floor living level into the hillside and to create a garden terrace for the second level. Steel beams set into the retaining walls perpendicular to the hillside are cantilevered off structural shear walls at the front of the site.

Lateral steel clear spans fifty feet between these beams creating a double cantilever at the leading edge of the house and allowing for uninterrupted views over Los Angeles. Front, side and rear elevations of the house slide open to erase all boundaries between indoors and out and connect the spaces to gardens on both levels.

Glass, in various renditions, is the primary wall enclosure material. There are forty—four sliding glass panels, each seven feet wide by ten feet high and configured to disappear into hidden pockets or to slide beyond the building perimeter. Deep overhangs serve as solar protection for the double pane glazing and become progressively larger as the main elevation of the building follows the hillside contours from Eastern to Southwestern exposure. This creates a microclimate which surrounds the building, creating inhabitable outdoor spaces while reducing cooling loads within. Every elevation of the house opens to capture the prevailing breezes to passively ventilate and cool the house. A vestibule at the lowest point of the house can be opened in conjunction with glass panels on the second floor to create a thermal chimney, distributing cool air throughout while extracting hot air.

Glass in the form of fixed clear plate panels, mirror plate walls and light gray mirror glass panels lend lightness to the interior spaces. These glass walls are visually counterweighted by sculptural, solid elements in the house. The fireplace is made of dry stacked granite, which continues as a vertical structural element from the living room floor through the second story. The main stair is charcoal concrete cantilevered from a structural steel tube. Service and secondary spaces are clad in floor to ceiling rift oak panels with flush concealed doors. Several interior walls

320 | North America

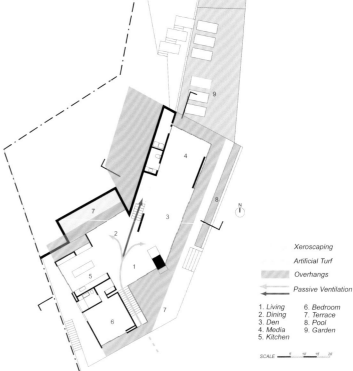

Xeroscaping
Artificial Turf
Overhangs
Passive Ventilation

1. Living
2. Dining
3. Den
4. Media
5. Kitchen
6. Bedroom
7. Terrace
8. Pool
9. Garden

SCALE 5 10 15 20

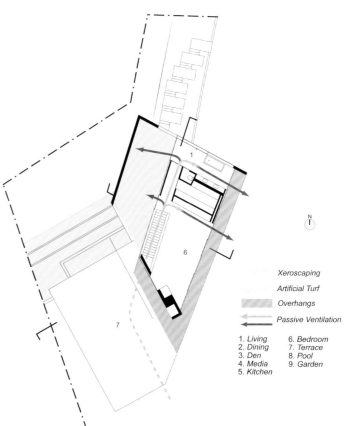

Xeroscaping
Artificial Turf
Overhangs
Passive Ventilation

1. Living
2. Dining
3. Den
4. Media
5. Kitchen
6. Bedroom
7. Terrace
8. Pool
9. Garden

are dark stucco, an exterior material that wraps inside the space. The use of cut pebble flooring throughout the house, decks and terraces continues the indoor—outdoor materiality, which is amplified when the glass walls slide away. The building finishes are few in number but applied in a multiplicity of ways throughout the project, furthering the experience of continuous open spaces from interior to exterior.

Set in a visible hillside area above Sunset Boulevard, the Openhouse appears as a simple folded line with recessed glass planes, a strong sculptural form at the scale of the site. The minimalist logic of the architecture is transformed by direct and indirect connections to the buildings' immediate environment. The perimeter landscaping is either indigenous or a drought—resistant xeriscape. An outdoor dining area implements artificial turf composed partly of recycled rubber. With the glass walls completely open the house becomes a platform defined by an abstract roof plane, a palette of natural materials, the hillside and the views.

Areopagus Residence

Architect: Paravant Architects, Saar Arquitectura
Location: Costa Rica
Photography: Julian Trejos

Areopagus is a building harmoniously integrated into the landscape of the Costa Rican mountains. The project was planned as a retirement residence for a client from Hollywood, California.

The building has two very different facades. The street facing facade is a massive concrete wall providing privacy for the owners. Openings in the solid wall—plane, frame controlled views of the surrounding mountains. These and other strategically placed openings provide for natural cooling and cross ventilation. The private south facing elevation, with large transparent glazing and deep roof overhangs completely opens up this side of the house to stunning views across the valley towards San Jose. The living room has a 16m long fully retractable corner glass—sliding wall allowing for true indoor—outdoor living. The adjacent pool with its vanishing edge enhances the already dramatic experience of this space. Additionally, the pool acts as a cooling pond reducing air temperature in the house. Through thoughtful design, sensitive consideration of the orientation of walls and openings, an awareness of existing site conditions and the use of passive strategies the building does not require a mechanical air conditioning system, drastically reducing the energy needed to maintain a comfortable living environment compared to buildings of a similar size and setting.

The project was designed and managed by Paravant Architects a German—American architectural firm with offices in Los Angeles, California, Charleston, South Carolina and Germany in collaboration with executive architect SAAR from Costa Rica. In considering issues of sustainability it was important for the architects to exhaust all passive measures before implementing strategically technical active systems. The active elements used in this house were roof installed solar and photovoltaic systems providing hot water and electricity. Additionally, an onsite micro waste—water—treatment plant was provided, as well as rain water collection for yard irrigation.

Echo House

Architect: Paul Kariouk
Location: Ottawa, Ontario, Canada
Photography: Photolux Studios (Christian Lalonde)

The starting point for this renovation is a modest Victorian home in poor condition, whose rooms with small windows and dark interior spaces are separated from one another, as is typical of homes built in an era when privacy was a cultural priority. In another gesture to Victorian public decorum, the arrangement of the existing interior spaces reinforced the antiquated ideal that work life and family life should be kept distinct.

Although the client required the renovated space to welcome work life in the home while continuing to maintain clear separations from their family life, it was also important to create a modern, bright space that, albeit small in size, would still appear spacious and visually connected.

Echo: noun , "a sympathetic or identical response, as to sentiments expressed; a lingering trace or effect".

That the house had scarcely been changed since it was originally constructed was both a virtue and a challenge that enabled design opportunities. The footprint of the house is small (approximately seven—hundred square feet). It was therefore not possible to create a loft—like setting on the ground floor that seemed simultaneously open and provided the required distinct work/living spaces. The house was re—envisaged as a vertical loft — an open, four—storey volume reaching from the basement to the ceiling of the new roof. The new main level and former basement level are opened to each other by a wide stair that highlights views to the home's original stone foundation walls. Hence, the former Victorian main living level, once segregated into four separate rooms, is now made open and spacious. The small, original windows are replaced with large windows both at the front and rear of the new parlour, visually extending that space into the front yard and the back yard, and, finally, enabling views from the back yard (all the way through the house) to views of the Rideau Canal.

The remaining spatial requirements include very private spaces: a study that could accommodate several thousand books, a den, and a master—bedroom suite. In order to achieve the seemingly paradoxical request for a

loft—like home but with spaces as private as the former Victorian ones, the study, den, and "book vault" are designed as distinct volumes suspended inside the larger, four—storey volume. Because these volumes "float" high up within the now—emptied shell of the original house, they achieve the required visual privacy from the parlour below and the street outside (despite the expanded areas of windows). Though these spaces are small, they are bright and airy and seem large since they all have visual access to both windows and other interior spaces of the home.

The very most private areas of the redesigned house (such as closets, toilets, and stairs) are arranged along the south wall of the house and are shielded by a three—storey hickory "modesty screen". At the top level, the master—bedroom suite cantilevers over the front facade and yard and also appears as a distinct, floating volume, and forms a canopy over the entry. In this way, the former attic space of the Victorian house is redesigned to provide for light and views where none existed before in the original home, and due to its elevated position, it does so while maintaining privacy. At the initial request of the clients, this renovation allows the values of a bygone era to be given voice in the current era.

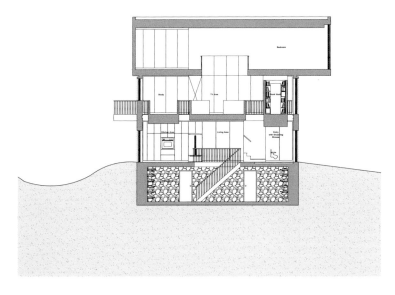

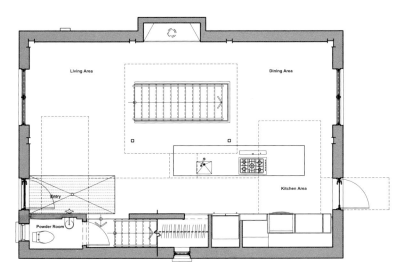

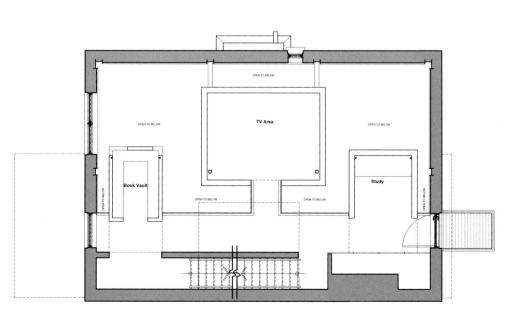
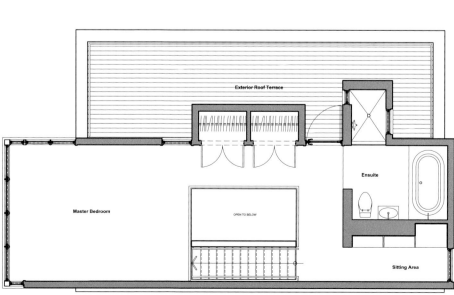

Hill — Maheux Cottage

Architect: Paul Kariouk
Location: Val—des—Monts, Québec, Canada
Photography: Photolux Studios (Christian Lalonde)

In this small, weekend and vacation retreat, the clients and their daughter seek to take refuge from the world. As such, the home is introverted, but the clients also desire that the house achieve a maximum connection to its beautiful, forested, lake—side site. Last, it is important to note that this home is built as a place the clients will keep for their entire lives and then pass on to their daughter; the clients, a couple each involved in art conservation, sought a home whose design would itself be conceived as a vessel for the conservation of the family memories that will continue to unfold here.

The design of the cottage is simple: two "bars" of living space that are joined by an elliptical loft hovering over the foyer and giving shelter to the entry below. One bar is private, containing bedrooms, bathrooms, and a storage; one bar is public, containing the kitchen, dining and living areas. The elliptical loft is the domain of the daughter.

Rather than "walling—in" the two ground—level volumes to achieve privacy, they are made with large expanses of glass and are sited at the edge of the property where vegetation is most dense. The underside of the loft volume and the fireplace surround are surfaced with a "quilt" of metal plates, including copper and zinc printing plates that the clients received from a printmaker friend. Many of the plates are etched with landscapes from the original printmaker, but many are etched with works created by the couple and their daughter. There are, however, many yet—unetched plates that can be removed, worked by the clients and their friends visiting the cottage, and then reinstalled. In this way, the house keeps a record of its past.

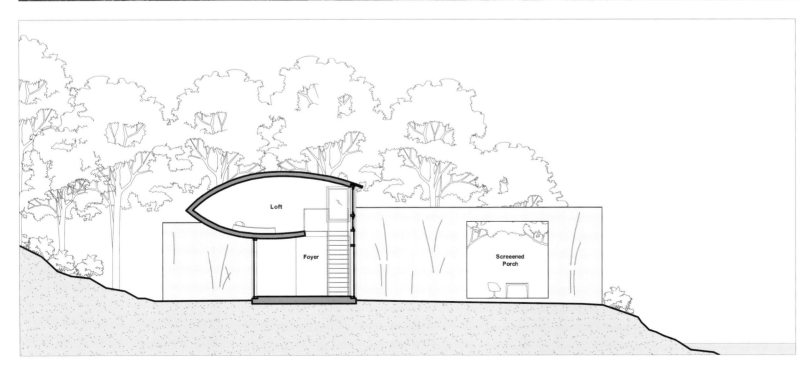

Hurteau—Miller Cottage

Architect: Paul Kariouk
Location: Val—des—Monts, Québec, Canada
Photography: Photolux Studios (Christian Lalonde)

This cottage is designed for a couple with a young son and will serve as a center—point for the gatherings of a greatly extended family and will one day become the family's primary residence. The challenge is to implement an architectural strategy that breaks down very large programmatic requirements and, as a result, the seeming bigness of the proposed home. The spatial system therefore is conceived as a porous and open ground—floor that supports a solid, "private" volume above. All of the primary living and communal functions, such as the indoor and outdoor living rooms, kitchen and eating area, and sauna open either literally or volumetrically to the surrounding lakeshore. The master bedroom suite is also placed on the ground floor since the house is meant to be a home for a lifetime and issues of future accessibility are thus addressed. The upper volume, a heavier wooden box clad of Western Red cedar, is cradled by the seemingly delicate armature below and houses spaces that are introverted, namely the son's bedroom and guest sleeping areas. This dropping of the second—floor volume into the ground—floor level is meant to create spatial interest and intimacy throughout the large, communal areas of the interior while minimizing the house's overall size from the exterior.

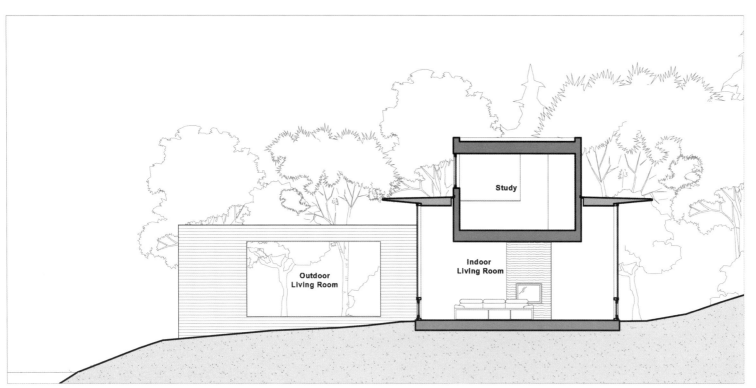

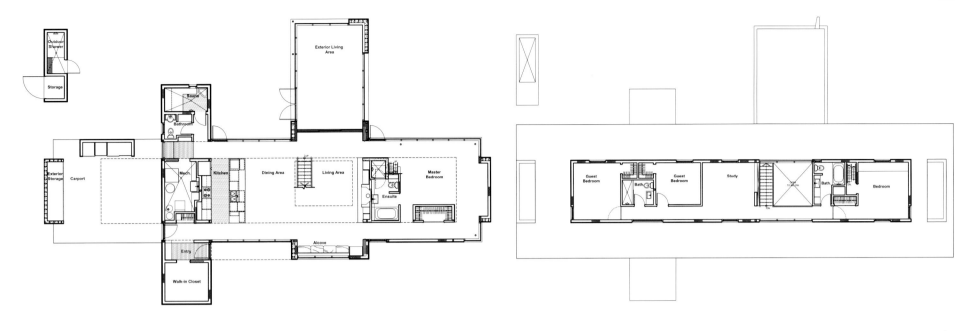

Pricila

Architect: Martin Gomez & Gonzalo Veloso
Location: Nuñez, Buenos Aires, Argentina
Area: 1,000 m²
Photography: Alejandro Mezza

Pricila is a bet to the growth of a district resurgence. In Buenos Aires, "river" is at its oment a residential district that ended up discredited due to the soccer mega stadium and its recitals. With the resugente of the nuñes neighborhood, river (located just steps away) started to recover its splendor in hand with new institutions, projects, and high—class restaurants.

With this in mind the architects create a house that nowadays is perceived as a satellite, since the bordering buildings show this passage between yesterday and tomorrow. Pricila represents the desire of future. With a concrete structure, which is shown in the facade, it invites its neighbors to guess what's in its interior.

With 1,000 m² constructed in a palace structure, with ceilings of 3 meters high, and spectacular dimensions, Pricila surprises everyone who sees it in every corner.

The materials chosen are of great nobility such as lapacho wood for all its inner floors, or marble for all the exterior including the pool entirely sided in marble. The baths show these same sensations, with marble siding from floor to ceiling. A circular stair acts as the central focal point made of concrete, wood and stainless steel, that ends up with a circular skylight that crowns the staircase. The house opens completely to its interior, highlighting its tropical garden, like an oasis in the middle of the city, and ends up closing to the exterior with wooden shutters, and a huge concrete frame that encloses the first floor box. Following the guidelines of the studio, the project always accompanies the lifestyle of its occupants, this house invites them to enjoy the outside, with barbecues, outdoor chimneys, jacuzzis, pools, and even practice shooting spaces.

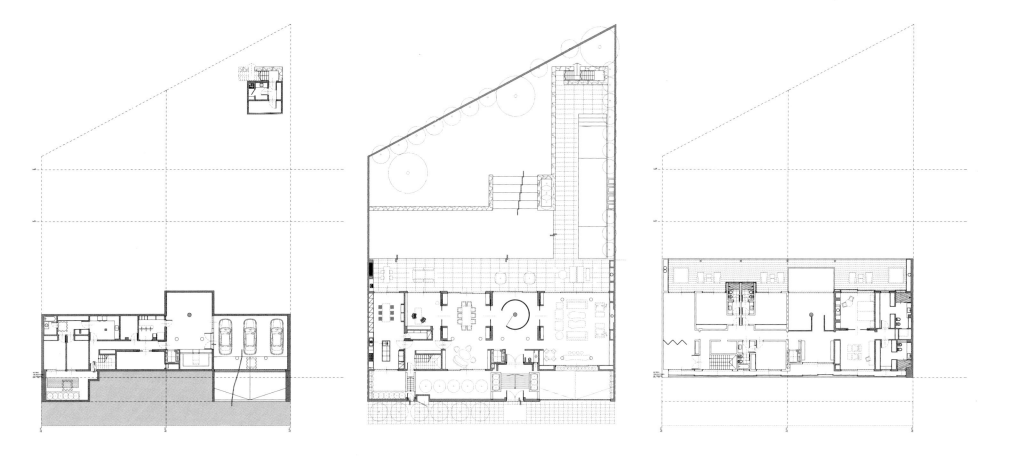

Ribbon House

Architect: G2 Estudio
Location: Rio Negro, Argentina
Area: 396 m²
Photography: Laila Sartoni

Linking distant lands, this time G2 Estudio was hired by two families from Tahiti — French Polynesia, partners in the adventure to create a holiday house in San Carlos de Bariloche — Patagonia Argentina, the one should have a wide integrated space for leisure and recreation, two master rooms, two bedrooms for the children, and all the necessary equipment to a holiday house...

Like the current residence style of the owners (on an island), the choice of the area where the house would be implanted also seems to be an island, but this time between two rivers, surrounded by a stunning landscape into the four cardinal points: Cerro Catedral to the West, Sierra Ventana to the South, the Golf to the North, and a canyon with a creek and lush vegetation to the East. This way the house should be sufficiently dislocated in the footprint composition to ensure that each volume can achieve a particular frame of the breathtaking nature...

The initial idea comes from the juxtaposition of volumes, each containing different functions, on one hand the social life and in the others the private life. When these volumes meet each other, mixing the geometry and the space, they generate dynamic routes between the activity and rest areas of the house, and getting in tension they experiment the transition between being supported on the rock to raise into the sky searching perfect visuals.

Is this way that the architects can appreciate an up—down experience link. The morphology and materials used, are thought to achieve the strong in fragile, the solid in ethereal, the supported in support, the dynamic in static, and vice versa. So the house is a search between the balance, juxtaposition, ribbon, viewing—point, vital tour, and hug.

To get the artistic expression and for reach the limits of the materials, the work was performed with two different systems that can reflect the idea of the project. The support would be reinforced concrete, bringing

it to its fullest potential in horizontal planes, vertical and lead off, for a seismically active area such as San Carlos de Bariloche, along with the stone as a heavy and rustic material in dialogue with the nearby mountains. The sustained would be of steel—frame for the outer shells, partitions, panels, sun visors, which would be clad in wood from the area taking advantage of its warmth and lightness. For the roof panels and folds, is used asphalt slate black color, creating color and texture contrasts.

The interiors are the result of the interpenetration of volumes compositional directions respecting convergences and materiality making a dramatic balance between the expression of forms, textures and visuals.

With Ribbon House G2 Estudio close a small cycle of evolution in the search for housing types, and launch into new spaces for architectural exploration, arguing that every expression of architecture should be unique and unrepeatable as the users are.

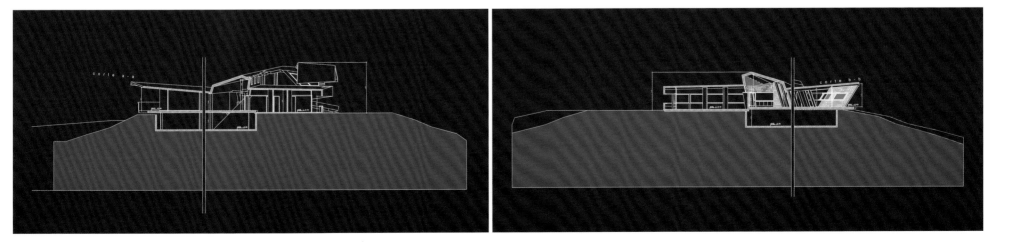

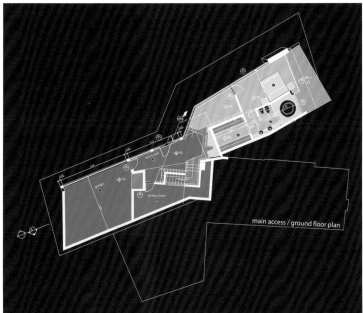
main access / ground floor plan

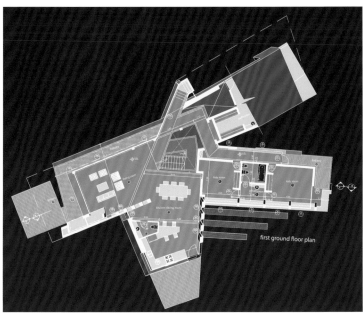
first ground floor plan

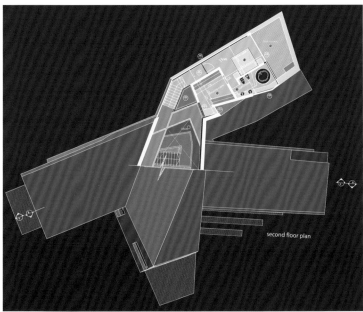
second floor plan

House L

Architect: Mathias Klotz
Location: Buenos Aires, Argentina
Site Area: 1,330 m²
Photography: Rolan Halbe

House L is a single family house located in Olivos, an old neighborhood in Buenos Aires. The site is rectangular (24x54 meters) and presents series of trees randomly distributed in the whole area.

The project divides the program around the garden, thus allowing the constant interactivity with it in a contemplative way. The program is organized in series of squares linked by a linear circulation. Moreover, the interior spaces are linked to the exterior in more than one face.

The materials used are concrete, travertine, steel and wood, which are linked to the landscape and achieve a fluid dialogue between the interior and the exterior.

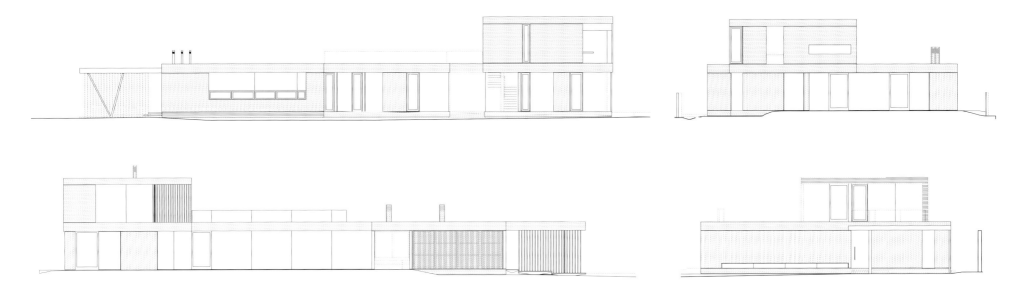

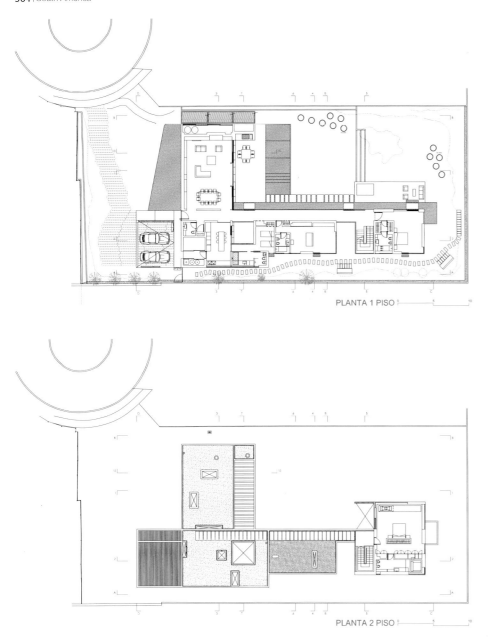

PLANTA 1 PISO

PLANTA 2 PISO

Casa Maritimo

Architect: Seferin Arquitectura
Location: Rio Grande do Sul, Brazil
Area: 360 m²

The Project was commissioned for the Casa & Cia praia 2012, an architecture and decoration event that occured during the summer on the coast of Rio Grande do Sul, Brazil.

Designed and built in the year of 2011, the 360 m² residence occupies a corner lot of a particular condominium in Tramandaí beach.

The shape of the residence is very clear, divided in two distinct blocks: a two story linear brick volume and a ground level glazed block.

The living room, dining room and the gourmet area, are fully integrated. Glazed, with glass sliding doors, this block is surrounded by a balcony deck high ground. A slab of concrete makes the cover of this volume, sheltering in a second floor terrace with magnificent view. The brick fireplace is outstanding volume of the environment, and can also be used outdoors on the porch.

On the ground floor, the longitudinal block houses all the service of the house, facing an external service circulation with their own access, and a guest suite with private balcony facing the garden back of the house. On the second floor, three suites are connected by a main circulation that opens on to the terrace overlooking the lagoon. The master suite sets forward on the floor, making the coverage for the car. A large window with view over the area of preservation of the condominium is the focal point of the suite.

The decoration of the social area seeks the same language of the house architecture. The materials used externally — brick, wood and cement — are also used internally, bringing the same elements of the facades into the house. These materials allow the house to merge with the landscape. The selection of furniture seeks a balance between rustic and contemporary, mixing design pieces and elements with natural materials.

Residência Belvedere

Architect: Anastasia Arquitetos
Location: Belo Horizonte, Brazil
Site Area: 450 m²

The 370 m² residence distributed in two levels is situated in a manly residential neighborhood in the city of Belo Horizonte, on a 450 m² flat site. The architectural approach seeks to privilege the maximum integration of external and internal areas, mixing up their boundaries, and, then amplifying the feeling of wideness.

Due to the reduced size of the site, residual and crossing spaces are practically left out (for example, there is no entrance hall, in behalf of a visual permeability with the entrance garden, achieved through large pivotal doors in the facade).

The floor plan is rectangular and compact, stretching till the site's sidelines. The rooms are illuminated by large doors front and back facades and also by matted glass locking (u—glass that acts as a good thermal insulation due to the existence of an air layer between the glass sheets) between the lagged cover slabs. A glass cover over a concrete pergola complements the illumination through an indoor garden. Therefore, the house is flooded by zenithal and indirect natural light that besides avoiding artificial lighting during the day, also avoids excessive heat from direct sunlight. The prevailing wind comes from the street, thus the entering doors work as regulators of wind speed. Totally opened in the summer, they enhance cross ventilation, or closed in the winter, or even semi opened if little ventilation is desired.

The residence was established in the street level, one meter above natural ground, in order to avoid unevenness and improve accessibility of the social areas. And, it also let the house more protected from the soil moisture.

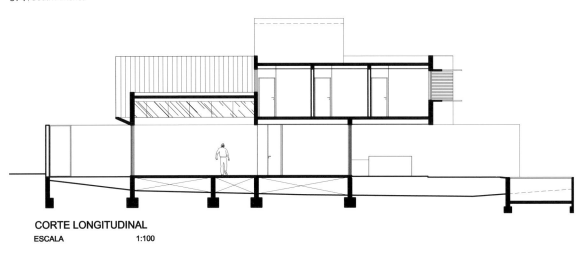

CORTE LONGITUDINAL
ESCALA 1:100

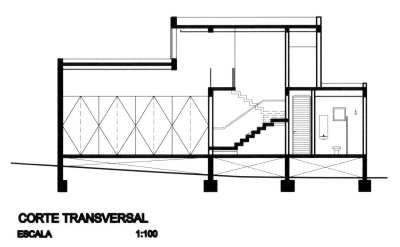

CORTE TRANSVERSAL
ESCALA 1:100

It is important to remind that one of the reasons for the implantation of compact field, reducing its footprint, is to increase the permeability of the ground, something really needed in our cities.

Solar collectors (that meet the house and the pool) occupy the most of the cover slab which prevented the use of this area initially contemplated.

Due to the large spans desired, supported by few points of foundation, and also to the large porch swing, the upper walls are concrete beams built by ripped forms of wood left apparent. Its aesthetics comes from a structural option, such that it is not decorative. This structural gymnastics is important, as the support pillars on the porch would be contrary to the intention of integration between interior/exterior as desired.

The result is a lightweighted residence (despite its aesthetics of exposed concrete), lighted and ventiladed, with pleasant and proportional spaces, that puts into to practice the initial desire to the best possible use of external area.

Vila Castela Residence

Architect: Anastasia Arquitetos
Location: Nova Lima, Minas Gerais, Brazil
Floor Area: 650 m²
Photography: Jomar Bragança

Built in a sloped site (30 degrees) in the city of Nova Lima, Brazil, the house uses dramatic cantilevers to emphasize the extremity of its position.

The architects have chosen this concept not only for aesthetic reasons, but above all to reduce the interference of the building mass in the topography, keeping the site as natural as possible.

As the architects placed the ground level of the residence 7 meters below the street , They are able to preserve the pedestrian view of the forest, at the same time keep the privacy of the owners, because the main apertures and the glass walls are oriented to the east, on the opposite side of the street. The urban impact of the residence is minimized, in benefit of the beautiful view of the woods.

As the climate of this region is very good, with an average temperature of 28° Celsius in the summer, and 16° Celsius in the winter, the right orientation of the doors and windows prevents the use of artificial climate. Solar voltaic cells are placed on the roof.

The constructed area is 650 sq meters, divided into three floors: the basement, where the owners can enjoy leisure facilities such as sauna, Jacuzzi, and an wine cellar, the ground floor, where the living area and the kitchen, integrated to the outdoor swimming pool and external terrace, making it the centre of the house, and the first floor, where the occupants can obtain privacy in the bedrooms.

The form is generated by the engineering of the concrete structure, which is robust and sculptural and, at the same time, light and contextualized with the surroundings. The concrete maintains its texture, and the

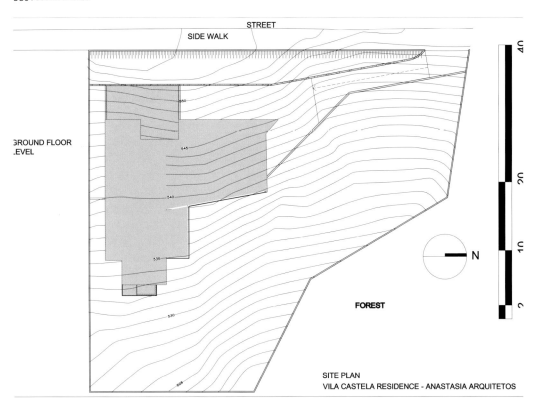

SITE PLAN
VILA CASTELA RESIDENCE - ANASTASIA ARQUITETOS

masonry is painted terracotta, for low maintenance reasons. (the ground has a red dust, iron ore dust).

The designers use as few columns as possible in order to preserve the existing site. Unsurprisingly, given the exquisite surroundings, the largest proportions of the building face outwards down the hill with views of the forest. Glazed elevations make the most of these views and also of the sunrise to the east.

This project is based on four main concepts: little interference in the site, better use of natural resources, integration to the surroundings, and a generous urban presence.

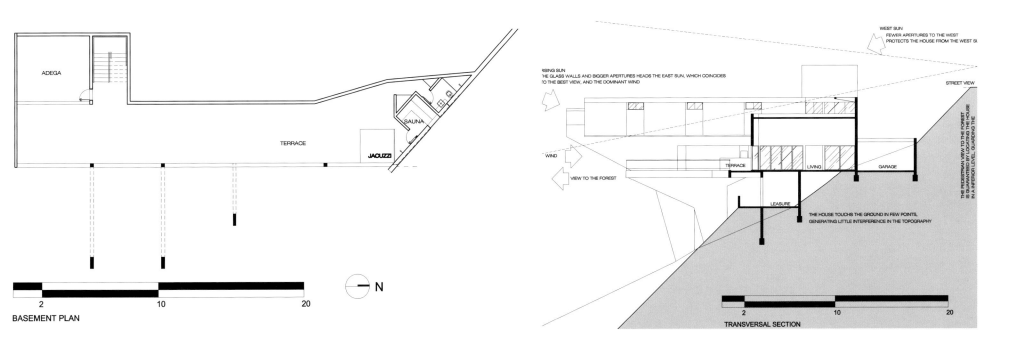

BASEMENT PLAN

TRANSVERSAL SECTION

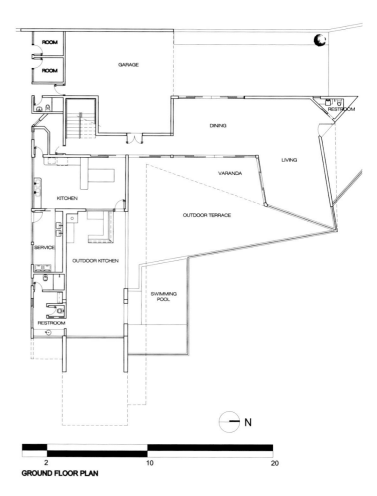

GROUND FLOOR PLAN

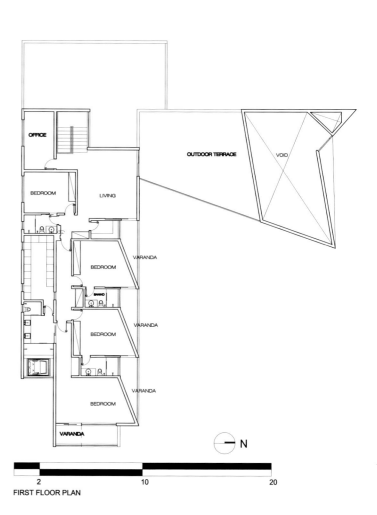

FIRST FLOOR PLAN

Manifesto House

Architect: James & Mau Architecture
Location: Curacavi, Chile
Built Area: 160 m²
Photography: Antonio Corcuera

The Manifesto House represents the Infiniski concept and its potential: bioclimatic design, recycled, reused materials, non polluting constructive systems, integration of renewable energy.

The project relies on a bioclimatic architecture adapting the form and positioning of the house to its energetic needs. The project is based on a prefabricated and modular design allowing a cheaper and faster constructive method. This modular system also allows thinking the coherence of the house with possible future modifications or enlargements in order to adapt easily to the evolving needs of the client.

The house, of 160 m² is divided in two levels and uses 3 recycled maritime containers as structure. A container cut in two parts on the first level is used as the support structure for the containers on the second level. This structure in the form of a bridge creates an extra space in between the container structures, isolated with thermo glass panels. As a consequence with only 90 m² worth of container, the project generates a total 160 m², maximizing and reducing significantly the use of extra building materials. This structure in the form of a bridge, responds to the bioclimatic needs of the house—Form follows Energy—and offers an effective natural ventilation system. It also helps to take full advantage of the house's natural surroundings, natural light and landscape views.

As if it had a second skin, the house "dresses and undresses" itself, thanks to ventilated external solar covers on walls and roof, depending on its need for natural solar heating. The house uses two types of covers or "skin": wooden panels coming from sustainable forests on one side and recycled mobile pallets on the other. The pallets can open themselves in winter to allow the sun to heat the metal surface of the container walls and close themselves in summer to protect the house from the heat. This skin also serves as an exterior esthetic finishing helping the house to better integrate in its environment.

Both exterior and interior use up to 85% of recycled, reused and eco—friendly materials: recycled cellulose and cork for insulation, recycled aluminum, iron and wood, noble wood coming from sustainable forests, ecological painting, eco—label ceramics. Thanks to its bioclimatic design and to the installation of alternative energy systems the house achieves 70% autonomy.

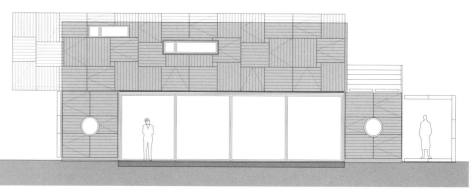

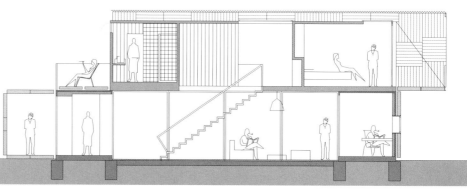

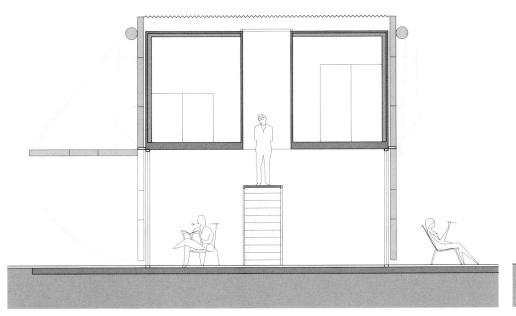
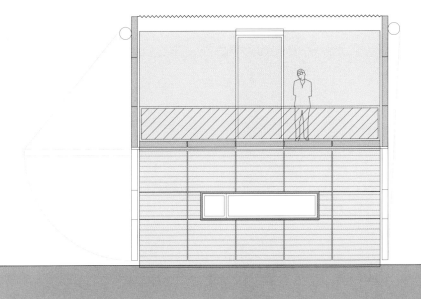

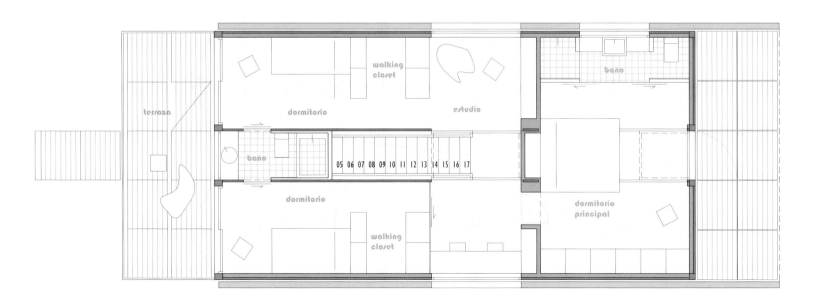

House Playa El Golf H4

Architect: RRMR Arquitectos
Location: Cañete, Lima'Peru
Land Area: 330 m²
Photography: Elsa Ramirez

The housing unit is a temporary summer home developed in the coastal desert of Lima, 95Km from the city. The stretch of land stands on a moderate slope in a privileged location with a frontal view to a golf course and a lateral view to the sea.

As a starting point, stands a large, comfortable terrace ending in a longitudinal swimming pool, which represents the main living space. On both sides of the plot, green areas may be found, which allow for natural illumination on the lower levels. These green areas define a lighter upper volume, which when separated from the base, takes the stage as the central element of the composition. This volume was given a starring role by working it in a closed way and by giving it a light, clearly defined aesthetic.

The sun sets towards the back of the lot, which is used in the solution to provide the necessary shade to the deck by extending the volume's upper edge, making it more distinctive and dynamic. The suspended main volume then shows a notorious diagonal that may be seen front the deck and from outside the plot. This resulting diagonal is taken as a composition gesture that is repeated as an inclined plane in the entrance, in the two stairways and in the irregular openings in the side facade.

Another element which incorporates the diagonal aesthetic is the perimeter walls. Along with providing privacy to the lower level, they bring together into the composition the suspended upper volume and the lower frame of the secondary bedrooms. This ends up unifying smoothly and continuously the unit's three levels into one single dynamic element.

The materials used, the walls painted in white, and the exposed concrete walls and veneered floors in gray, give the interior a neutral atmosphere, carefully illuminated, which allows to appreciate the use of space

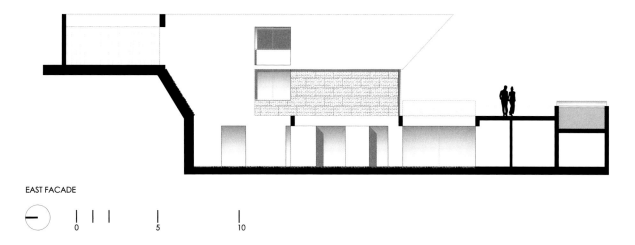

EAST FACADE

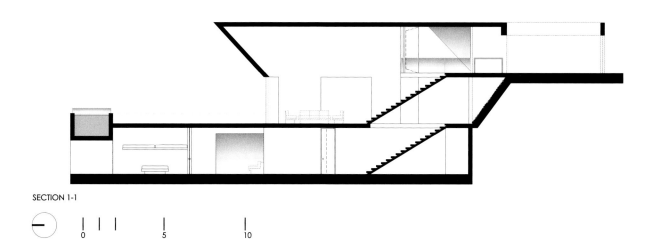

SECTION 1-1

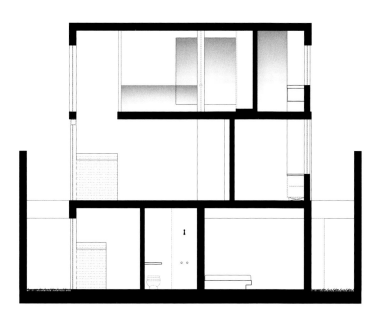

SECTION 2-2

1. ENTRANCE
2. PARKING
3. STOCK
4. BRIDGE
5. MASTER BEDROOM
6. BATHROOM MAIN
7. LIVING ROOM
8. DINING ROOM
9. SH. VISITAS
10. KITCHEN
11. TERRACE
12. BBQ
13. POOL
14. LAUNDRY
15. TENDAL
16. SERVICE BEDROOM
17. SERVICE BATHROOM
18. BEDROOM
19. BATHROOM BEDROOM
20. ROOM

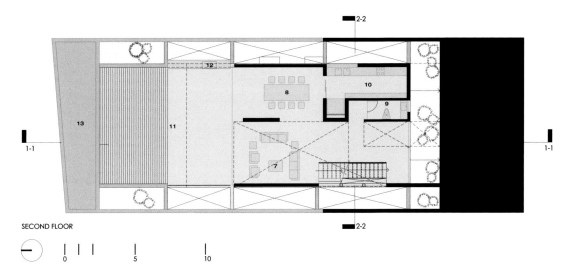

SECOND FLOOR

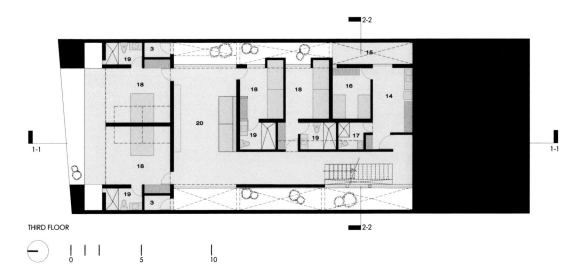

THIRD FLOOR

and the fluidity in the path between the different ambiances.

Access to the house is from the more elevated part in the back, and the layout considers three levels: the top gives way to vehicular and pedestrian access. The entrance is from the side with a tangential view to the suspended volume. An open courtyard leads to the hall of the house, and from this point, stairs go down to the social area, a bridge gives access to the master bedroom. The social area, pool and kitchen may be found at the mid—level. These areas are more open and are designed to connect with the landscape. The secondary bedrooms, family room and service area are found in the lower level. Two of the bedrooms have a frontal view, while the other areas are lit through the lateral openings.

The extensive program of this housing unit includes five bedrooms, a living room, a dining room, terraces, a swimming pool, a family room, a kitchen, two car garages, a storage area, a laundry area and a service bedroom and bathrooms.

Traditional building methods are followed, using a concrete structure and brick walls, plastered and painted.

Casa Playa Las Palmeras

Architect: RRMR Arquitectos
Location: Cañete, Lima, Peru
Land Area: 164 m²
Photography: Elsa Ramirez, Roberto Riofrio

The design is proposed in a compact stretch of land as a part of a residential development located in the deserted coast on the outskirt of Lima. This development is 120km away from Lima City.

The house benefits from the increasing slope to create a sculptural and aesthetic volume. The plot is located besides an open park. Despite the compact character one can see the lightness and dynamic properties of the volume, characteristics that strengthen its relation with the public park.

The monolithic and massive volume chosen as a starting point and worked as a big carved block searches the contrast in a duality introspection and permeability. This position prioresses the views towards the ocean and the park but closed to the immediate context.

Consequently, special attention was given to inner space and its visual to the environment. There are three visual axes that relate to different levels and its directionality with exterior. The first axis is horizontal from the main entrance and steered to the park. Perpendicular to it is the second axe which is vertical looking towards the sky and where circulation is developed vertically. The third one directs the view from the social area towards the ocean. These three axes merge in a single space that integrates not only the most important visuals and circulations but the social and intimate areas.

In this atmosphere, integration of conceptual visuals is achieved, as well as the functional integration of all areas through a spiral staircase bonding the second and third floor, leaving the first floor to parking and services areas. Light enters through visual frames covering all the house. Despite the house is compact, this effect gives the sensation of lightness and amplitude. The rest of the building closes to the environment giving it more privacy and avoiding neighbors' registry. The interior—exterior relation is emphasized through these visual frames.

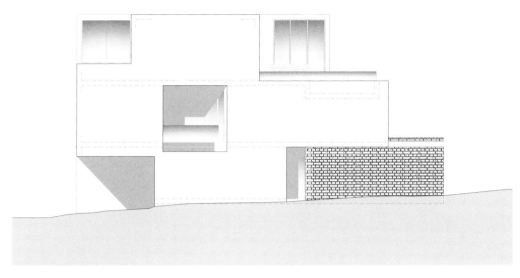

ELEVACION NORTE

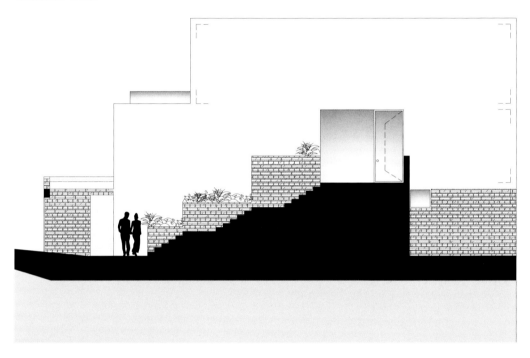

CORTE 1 - 1

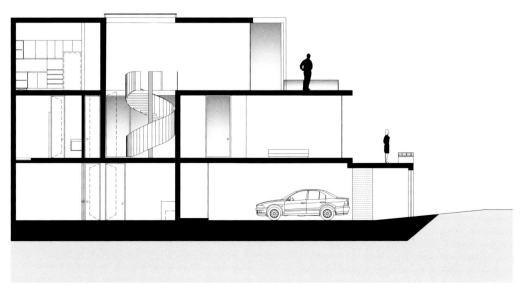

CORTE 2 - 2

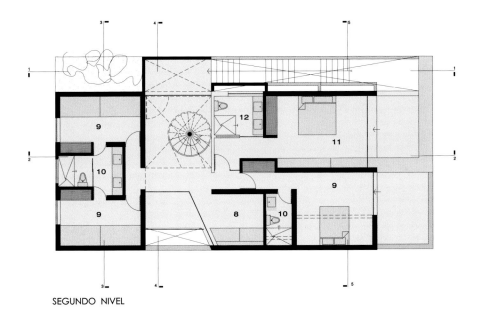
SEGUNDO NIVEL

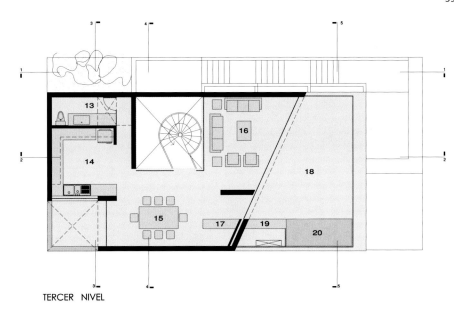
TERCER NIVEL

The materials used, the walls painted in white, the stone walls at the bottom and the grey floors seek to give a solid image in the exterior and a neutral atmosphere in the interior empathizing the strong colors of nature outside the house such as: ocean, park, sky and mountains. The color inside the house is concentrated in the spiral staircase.

The architects chose to leave the social areas in the third floor, to take advantage of the ocean view over the first row terraces. In the first floor areas were identified which required a lot less illumination such as service and parking areas.

The plot of the house is compact and functional. One enters by a lateral staircase that leads straight to the second floor. There one immediately discovers the principal perforation in the house volume which relates with the park in the other corner of the house. From this point one can access the four bedrooms and the family room. The spiral staircase connects this level with the top level where the social areas are located such as dining room, terrace, pool, living room, all of them with a direct view to the ocean. In the fist floor are the services areas and a parking for four cars.

Traditional building methods are followed using a concrete structure and brick walls with white rendering.

Black and Red

Architect: Martin Gomez
Location: Jose Ignacio, Punta Del Este, Uruguay
Area: 340 m²
Photography: Ezequiel Escalante

This house was planned for a family with kids of different ages, who therefore are at different stages of life, for that matter the architects decided to create a house that would articulate 3 different volumes connected by a concrete galleria. Each of these volumes is completely independent attached to an open galleria that finally opens to a big swimming pool and a generous wooden deck as their daily meeting point.

This galleria is enclosed between two concrete fireplaces, one for cooking and the other one for comfort, consequently creating a generous interior space below the galleria.

This house of modern lines brings simplicity to the way its occupants spend their summer.

These three volumes contain master bedroom and bathroom, living room, kitchen and toilette, kids bedroom, bathrooms and playroom.

The materials used are concrete, red colored and black colored wood, trying to play with the different lightning the sun offers in Punta Del Este.

Through this new system of living the owners discovered a completely different way to enjoy life, in a more relaxed and distended way, opposite to their life in the city where they live.

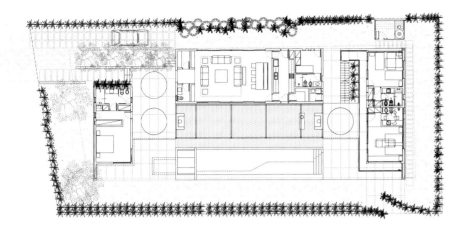
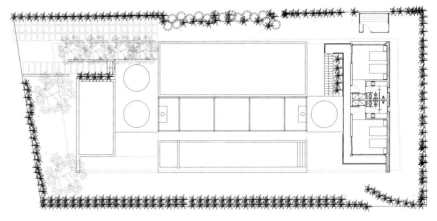

PLANTA ALTA ESC:1/100

Muriado

Architect: Martin Gomez
Location: Manantiales, Punta Del Este, Uruguay
Area: 120 m^2

This house was conceived as a rental summer house, located next to the owners real house.

it was designed after 2 concepts, the first one is to take advantage as much as possible of the built area, so as to accommodate as many people as possible, in this case 4 rooms, and the second concept is to try to appreciate the sea views from every angle.

To achieve this the architects elevat the social area from the first floor, perch on concrete stilts, surrounded by a deck. It's a fusion between wood, concrete and glass. It is an awkard plot so the project accompanies its triangle shape.

The project is conformed of 2 blocks, one of concrete that contains the private areas (rooms), and the other with wood siding one on the second floor, with the former being conneeted by a concrete pergola to this block.

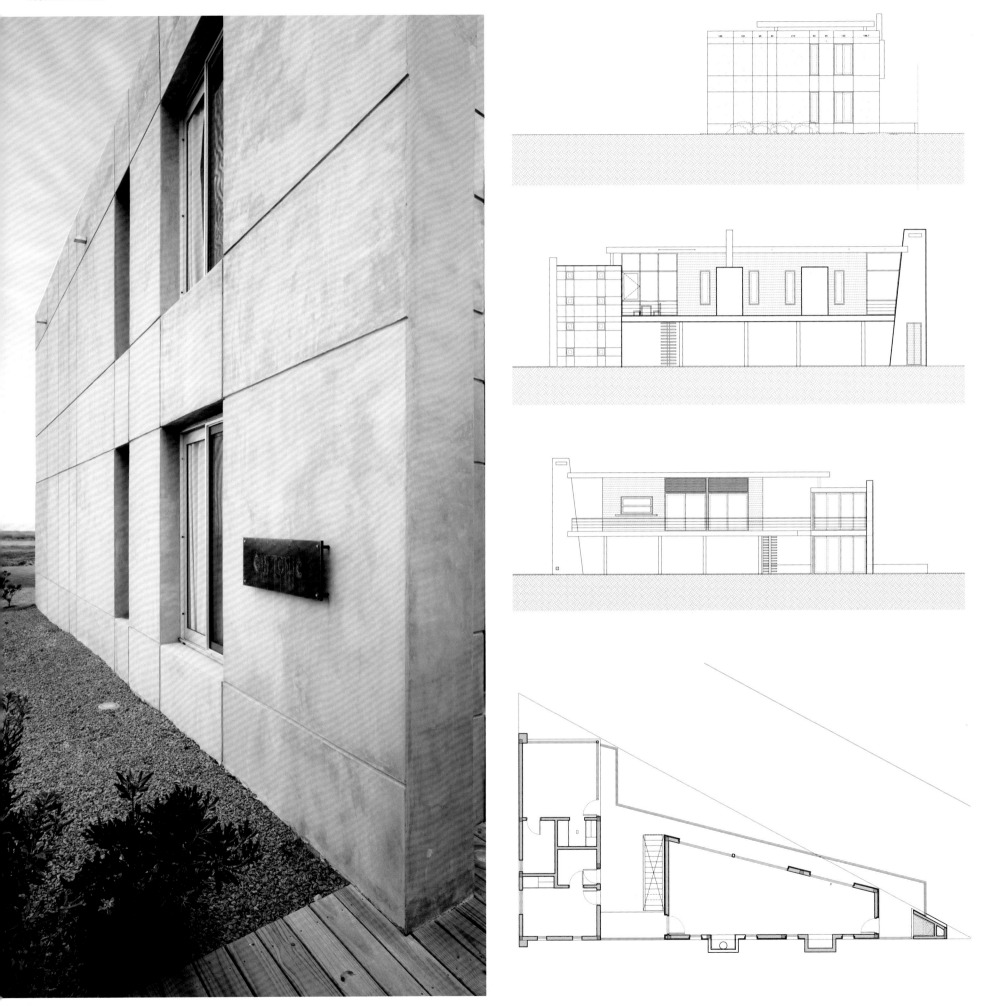

Psicomagia

Architect: Martin Gomez
Location: Punta Piedras, Punta Del Este, Uruguay
Area: 300 m²
Photography: Ezequiel Escalante

This house is projected with two volumes. the first one overlooking the Ruta 10, which contains the service area. and the second one overlooking the garden, pool and sea. The ground floor has 3 bedrooms and the first floor serves as the social area with an amazing view to the Punta Piedras beach.

The project has the two volumes facing each other connecting the ground floor by a terrace as a transition area, which creates a straight dialogue between each other. At the same time these two volumes differentiate each other. The service area is a closed and heavy block and the social area is more open and transparent opening itself to the most important views to the terrace and the sea.

In the first floor several pergolas play with the social volume creating various outdoor situations that invite its habitats to enjoy al fresco.

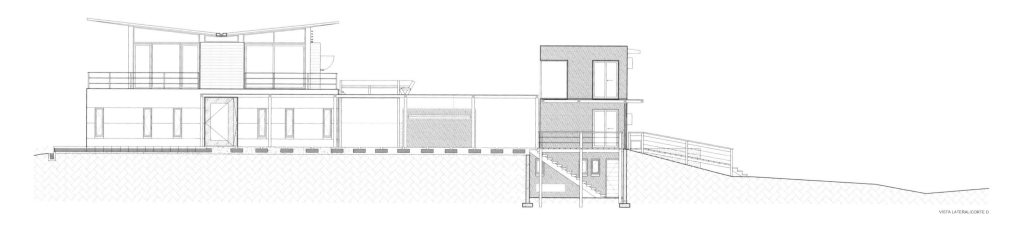

VISTA LATERAL/CORTE D

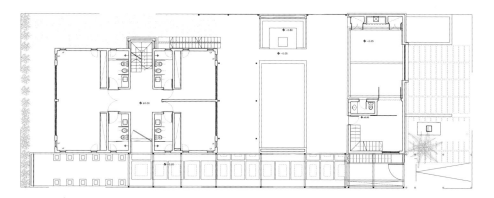
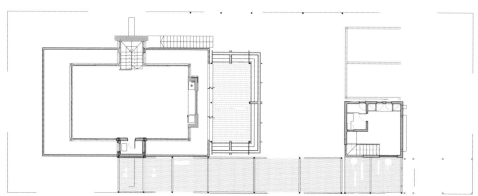

Villa Ketty

Architect: Martin Gomez
Location: El Chorro, Punta Del Este, Uruguay
Area: 380 m²
Photography: Ezequiel Escalante

Villa Ketty is a combination of glass, stones, concrete and white painted walls. The project articulates three levels, flanked by an eternal stair and an internal one. The social area is located on the upper floor while the suites are on the ground floor.

One of the most beautiful details is hidden in the pool, which contains a series of circular openings that allow one to see te subaqueous world from the inner circulation of the house. A true spectacle.

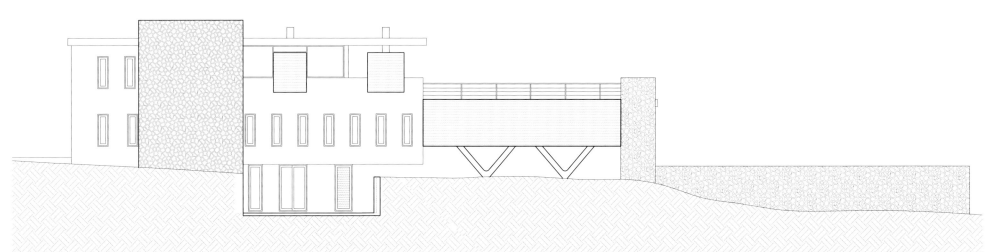

Casa Carrara House

Architect: Andrés Remy Arquitectos
Location: Pilar, Prov. de Buenos Aires, Argentina
Area: 600 m²
Photography: Alejandro Peral

Located on an irregular lot, the house sits at the back of the lot and is parallel to one of the streets to open the best orientation and capture the best views. The idea of this journey was to discover the entrance as we follow the exterior stone wall. The rustic and crafted stone defines and separates the entry zones from the living spaces and is inside and outside, proposing a counterpoint to the pure white that dominates the inside of the house.

A blind and evocative entrance makes a strong impression on the house. The white of the carrara marble dominates the interior architecture. With the white walls and ceilings, the house appears to arise from within the water. The touches of color are used for small details and decorative objects, dominating the white color and the turquoise of the water.

The water that surrounds the house penetrates it in the form of the mirrored surface of the water whose novelty results in the interior cascade that emerges from the top floor and falls while painting reflections via a pane of glass. This mirror of water is reproduced outside blurring the boundaries between one and the other. Finally, emerging from the hall dispenser of the top floor, the glass cascade drains musically into and through the heart of the ground floor. These elements give the project the distinct mark of Remy—bold, creative and perhaps provocative, but always unique.

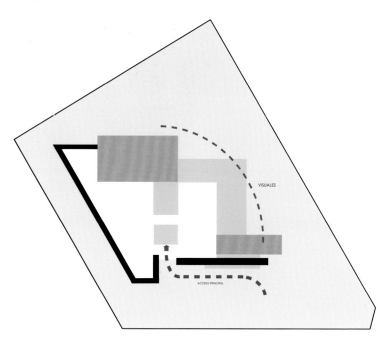
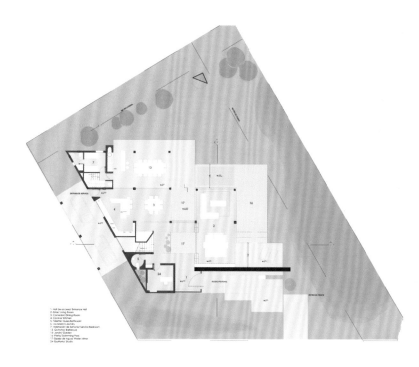

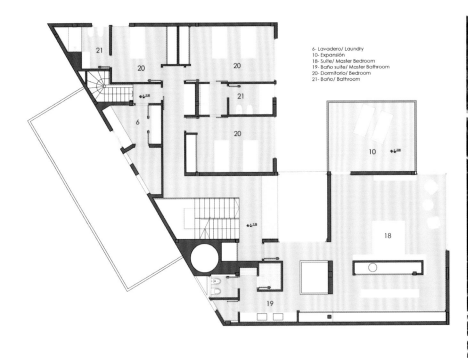

6- Lavadero/ Laundry
10- Expansión
18- Suite/ Master Bedroom
19- Baño suite/ Master Bathroom
20- Dormitorio/ Bedroom
21- Baño/ Bathroom

Cabo House

Architect: Andrés Remy Arquitectos
Location: Benavidez, Prov. de Buenos Aires, Argentina
Area: 400 m²
Photography: Alejandro Peral

The Project is located on a closed community on the north of Buenos Aires, on a lot with north orientation and surrounded by a lake on two of its sides, giving us amazing views.

The clients, an elder couple that no longer lives with its children, need a cosy, compact and overall functional home.

The house opens towards the back, matching the northern sun and the best views. On the east side, the house meets the lake through a semi covered space that functions as a garage and outdoor living area. This space allows the integration between the front garden with the back garden.

To the south, the house appears quite compact to preserve the owner's intimacy. The kitchen is on the side that takes the most of lake view beyond the street.

On the inside, there is a central double high green space — structuring axe of the house — that makes air circulation moving more freely, brings some indirect zenithal light to the different areas of the house and generates crossing views between them all. With the stairs, this space connects and separates visually serving and served functions. It creates this way a new landscape that organizes into a hierarchy the hall, the stairs and the suite.

The pool is strategically located on the edge of the lot, where it gets a lot of sun during all day and, at the same time, offers the best views thanks to the stairs which form like a bench that overlooks on nature. Moreover, it blends with the landscape's organic design and breaks up with strait shapes of the house.

The public ground floor rises with the terrain — which has naturally differents levels — so that it creates an interrupted view to the lake from the inside.

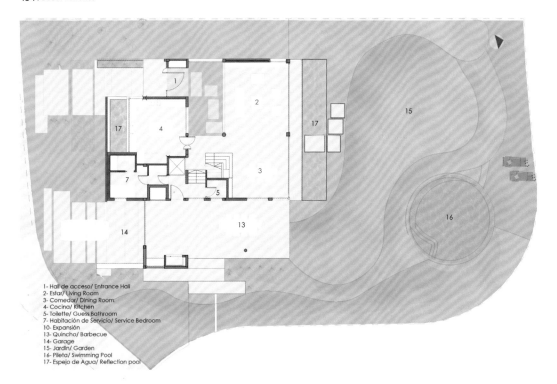

1- Hall de acceso/ Entrance Hall
2- Estar/ Living Room
3- Comedor/ Dining Room
4- Cocina/ Kitchen
5- Toilette/ Guess Bathroom
7- Habitación de Servicio/ Service Bedroom
10- Expansión
13- Quincho/ Barbecue
14- Garage
15- Jardín/ Garden
16- Pileta/ Swimming Pool
17- Espejo de Agua/ Reflection pool

1- Suite
2- Baño suite
3- Dormitorio
4- Baño
5- Escritorio
6- Lavadero
7- Depósito

A Clear Light in the City — Devoto House

Architect: Andres Remy Arquitectos.
Location: Villa Devoto, Capital Federal, Argentina
Area: 550m^2
Photography: Alejandro Peral

This project is a great challenge for the Studio, considering that it is going to take part of the urban structure. The lot of 18 x 24 meters is located in Devoto. It is between two existing buildings that threaten the project with the problem of isolation and view.

For this reason this project is leaning towards its neighbor, creating an introspective posture that creates the views from the inside rather than from the outside. An architecture that generates its own views.

In this manner the project is able to solve the basis of isolation and views from the garden. This has been a constant worry from the clients since the beginning of the Project.

The house of 500 m^2 was done for a young couple with kids. It is larger than the dimensions of given lot, threatening the space of the garden. For that reason, the first floor is given less program, leaving open the exterior spaces.

The living, kitchen and dining rooms expose the garden that penetrates all the volumes in the house. The spaces distort the conventional plans of an urban house. The stone wall guides the inhabitant within the house, therefore, the rough texture of the streets is softened by the water that flows down the stone wall, a walk that is colored by the different qualities, and the water that gives the space.

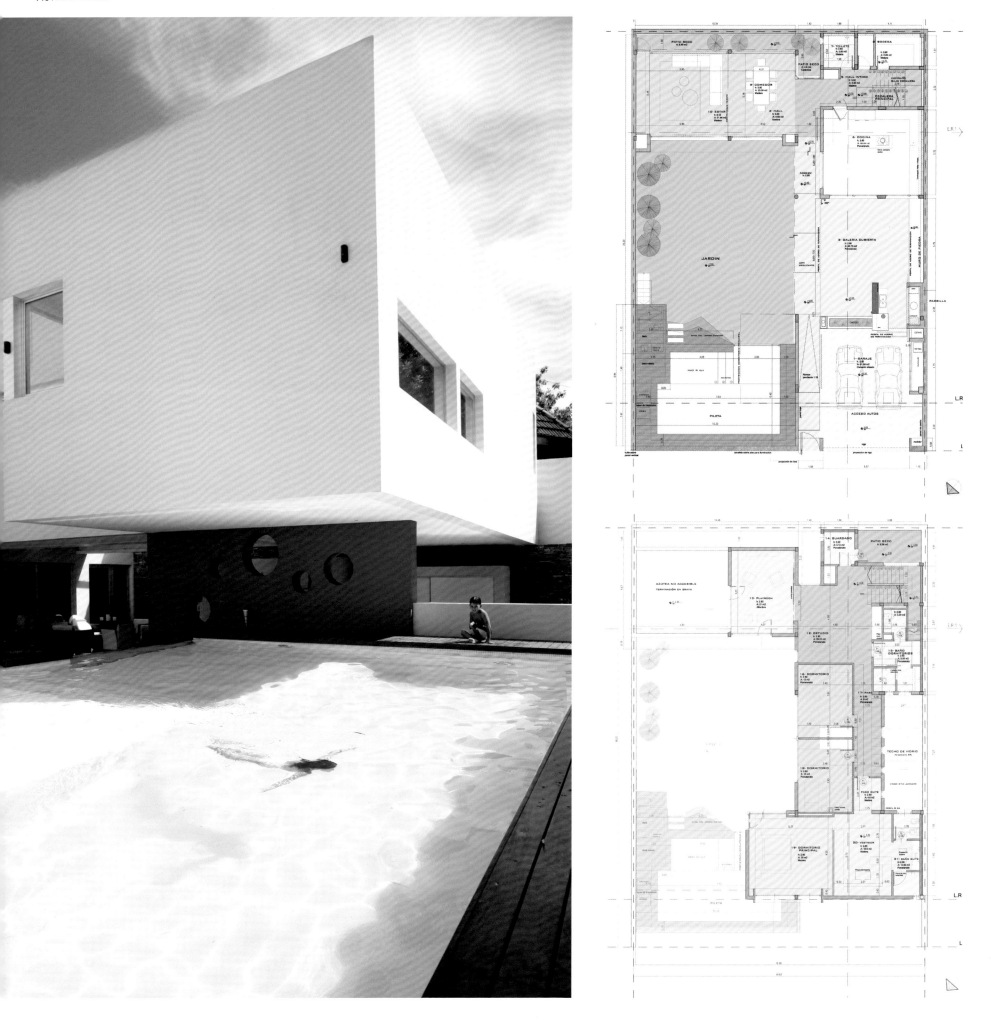

The green area between the living room and the neighbor guides the way to the entrance of the house. It is surrounded by the natural setting and different tones that the light of the sun gives to the space.

In this Project, the impact of the sun path is carefully studied, especially to place the swimming pool. One more time, water takes a big role in the creation of this house. This time the swimming pool is elevated from the ground creating a glass wall that allows views from within and the outside, creating views from different points in the house and focal point outside the house.

The house is formed vertically having a shifting effect. The bedrooms are located in the east overlooking the heart of the house, the garden. The last floor places a spa, having a jacuzzi, sauna and a gym with an exterior space, privately settled for the couple.

This Project goes beyond the rigid setting. It translates the urban setting into a natural setting subtlety. It pushes the boundaries of the conventional urban home to a spacious living where one has the best of both, the natural and the urban setting.

The Orchid

Architect: Andrés Remy Arquitectos.
Location: Pilar, Buenos Aires, Argentina
Area: 550 m²
Photography: Alejandro Peral.

The Orchid is a challenging project, committed by a young couple with two sons, enthusiastic about sustainable architecture. After consulting several architects, they give the commission to Andrés Remy, who have investigated these concepts in New York at Rafael Viñoly's Studio for four years.

Sustainability implies a lot of varieties, such as efficient and rational use of energy and water, natural ventilation and lighting, and low—environment—impact materials.

The house has the best orientation, possible thanks to the large lot. The concept come from the client's hobby, growing orchids. The house is based on the different parts of the orchid: the roots, the stem and the flower.

The sun rays impact in the interior of each room, are also studied, to determine the optimal depth to place the windows. This gives a unique volumetric outcome to the project. Taking advantage of sun rays in winter increases the interior temperature up to a comfortable level. The design includes glazed volumes with good thermal insulation and small windows in the worst orientation, such as the south facade.

The windows consist of aluminum frames supplied with thermal bridge breaker and double—hermetic—glass. A wide variety of insulating materials are also used, as well as water based paint and wall and roof air chambers.

In the lower floor, the convenient location of the opening windows allows natural air flow helping to decrease the humidity in the room. What's more, the big thresholds create a good distribution of fresh air.

All this, gives account of the complexity of the house, a glass and concrete flower, designed according to a program of needs and the client's concerns.

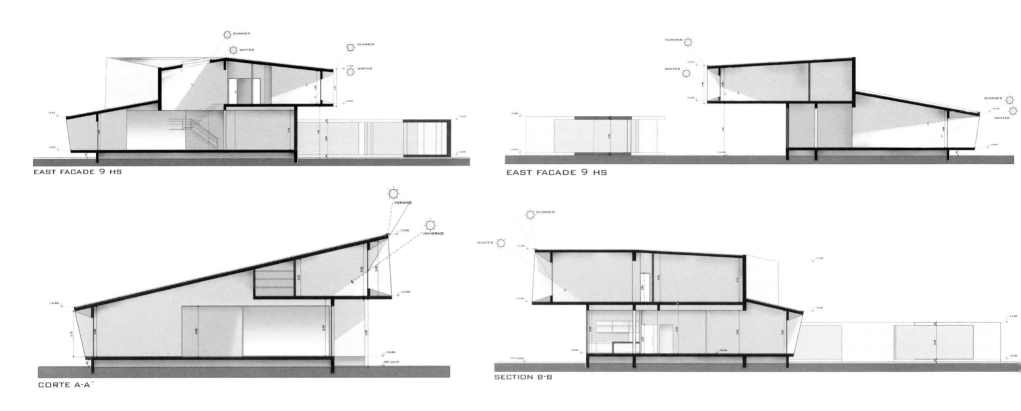

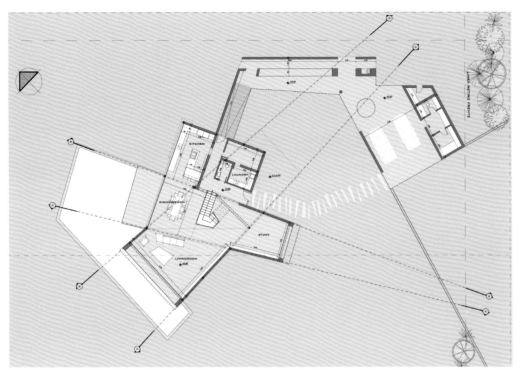
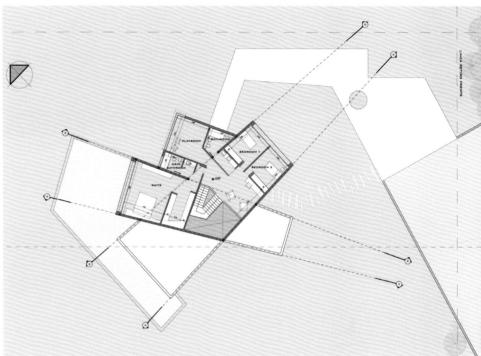